MECHANICAL D

By

JOHN S. REID, Sr., Mem.A.S.M.E.

Assistant Professor of Mechanical and Elementary Machine Drawing,
Armour Institute of Technology, Chicago, Ill.

British Library Cataloguing-in-Publication Data
A catalogue record for this book is available from the
British Library

Technical Drawing and Drafting

Technical drawing, also known as 'drafting' or 'draughting', is the act and discipline of composing plans that visually communicate how something functions or is to be constructed.

It is essential for communicating ideas in industry, architecture and engineering. The need for precise communication in the preparation of a functional document distinguishes technical drawing from the expressive drawing of the visual arts. Whereas artistic drawings are subjectively interpreted, with multiply determined meanings, technical drawings generally have only one intended meaning. To make the drawings easier to understand, practitioners use familiar symbols, perspectives, units of measurement, notation systems, visual styles, and page layout. Together, such conventions constitute a visual language, and help to ensure that the drawing is unambiguous and relatively easy to understand.

There are many methods of constructing a technical drawing, and most simple among them is a sketch. A sketch is a quickly executed, freehand drawing that is not intended as a finished work. In general, sketching is a quick way to record an idea for later use, and architects sketches in particular (in a very similar manner to fine artists) serve as a way to try out different ideas and establish a composition before undertaking more finished work. Architects drawings can also be used to convince clients of the merits of a design, to enable a building constructer to use them, and as a record of completed work. In a similar manner to engineering (and all other technical drawings), there is a set of conventions (i.e particular views, measurements, scales, and cross-referencing) that are utilised.

As opposed to free-sketching, technical drawings usually utilise various manuals and instruments. The basic drafting procedure is to place a piece of paper (or other material) on a smooth surface with right-angle corners and straight sides – typically a drawing board. A sliding straightedge known as a 'T-square' is then placed on one of the sides, allowing it to be slid across the side of the table, and over the surface of the paper. Parallel lines can be drawn simply by moving the T-square and running a pencil along the edge, as well as holding devices such as set squares or triangles. Other tools can be used to draw curves and circles, and primary among these are the compasses, used for drawing simple arcs and circles. Drafting templates are also utilised in cases where the drafter has to create recurring objects in a drawing – a massive time-saving development.

This basic drafting system requires an accurate table and constant attention to the positioning of the tools. A common error is to allow the triangles to push the top of the T-square down slightly, thereby throwing off all the angles. Even tasks as simple as drawing two angled lines meeting at a point require a number of moves of the T-square and triangles, and in general drafting this can be a time consuming process. In addition to the mastery of the mechanics of drawing lines, arcs, circles (and text) onto a piece of paper – the drafting effort requires a thorough understanding of geometry, trigonometry and spatial comprehension. In all cases, it demands precision and accuracy, and attention to detail.

Conventionally, drawings were made in ink on paper or a similar material, and any copies required had to be laboriously made by hand. The twentieth century saw a shift to drawing on tracing paper, so that mechanical copies could be run off efficiently. This was a substantial development in the drafting process – only eclipsed in the twenty-first century with 'computer-aided-drawing' systems (CAD). Although classical draftsmen and women are still in high demand, the mechanics of the drafting task have largely been automated and accelerated through the use of such systems. The development of the computer had a major impact on the methods used to design and create technical drawings, making manual drawing almost obsolete, and opening up new possibilities of form using organic shapes and complex geometry.

Today, there are two types of computer-aided design systems used for the production of technical drawings; two dimensions ('2D') and three dimensions ('3D'). 2D CAD systems such as AutoCAD or MicroStation have largely replaced the paper drawing discipline. Lines, circles, arcs and curves are all created within the software. It is down to the technical drawing skill of the user to produce the drawing – though this method does allow for the making of numerous revisions, and modifications of original designs. 3D CAD systems such as Autodesk Inventor or SolidWorks first produce the geometry of the part, and the technical drawing comes from user defined views of the part. This means there is little scope for error once the parameters have been set.

Buildings, Aircraft, ships and cars are now all modelled, assembled and checked in 3D before technical drawings are released for manufacture.

Technical drawing is a skill that is essential for so many industries and endeavours, allowing complex ideas and designs to become reality. It is hoped that the current reader enjoys this book on the subject.

PREFACE

A MECHANICAL DRAWING is used to convey precise information from one person to another.

A *patternmaker* must have a true drawing of an object, giving correct dimensions and instructions before he can make a pattern, from which the *foundryman* can make a rough casting.

The *machinist* must have a drawing from which he can obtain accurate information to enable him to take the rough casting and by slotting, planing, drilling, grinding, chipping or turning he can produce the finished article as designed by the draftsman.

Contractors, builders, architects, and engineers of all kinds, must have accurate drawings to enable them to produce satisfactory results in their work.

To do this it is essential that working drawings should be made according to certain principles and methods thoroughly understood by the man who makes the drawing and the man who uses it.

This volume on Mechanical Drawing is a fundamental course embodying all the theory, principles, and methods necessary to enable the student to make a practical working drawing.

Considering Mechanical Drawing as a language to convey thoughts and ideas; orthographic projection, which is a division of descriptive geometry, is its grammar and the foundation upon which is built all kinds of correct mechanical drawings.

This course is the result of twenty-five years of experience in teaching Mechanical Drawing and Elementary Machine Drafting (fifteen years at Cornell University and nearly eleven years at Armour Institute of Technology, besides Summer School and Evening Classes) and *twenty years* of *designing and drafting in practical work.*

In 1898, while at Cornell University, the writer produced a book on Mechanical Drawing entitled "A Course in Mechanical Drawing." In 1910 it was enlarged by adding short courses in Architectural Drawing, Sheet Metal Drafting and Elementary Machine Drawing. The present volume is offered, in the light of the above experience, as a complete course, not too long, preparatory to college work and as a foundation to practical drafting.

The divisions of the work included in the following course should be *standard* because they are all needed in the further development of draftsmanship. If any divisions should be emphasized more than others they are *freehand lettering, orthographic projection* and *isometrical drawing.*

This course is preparatory to a course in " Elementary Machine Drafting" by the writer soon to be issued from the press of John Wiley & Sons, Scientific Publishers, New York.

JOHN S. REID, SR.

ARMOUR INSTITUTE OF TECHNOLOGY,
Chicago, Ill., Feb., 1919.

CONTENTS

CHAPTER I

INTRODUCTION

THE following course is designed to train young men, who have satisfactorily completed the course in Mechanical Drawing, to become practical detail draftsmen and to lay a proper foundation for a future course in Machine Design.

It is the part of the detail draftsman to make the commercial working drawings of machine details for the use of the workmen in the shop.

This is usually done under the direction of the main draftsman who has charge of the design and construction of the whole machine.

In mechanical drawing, the student learned to make correct drawings of objects embodying the principles of orthographic projection or theory of drawing without much regard to the use of the objects drawn.

In machine drawing, however, more than this is required. The draftsman's motive in making a working drawing is to convey information by means of it to the men in the shop and, therefore, the drawing must be first *correct*, second, it must be made in as short a time as possible consistent with correctness and third, it must be as neat and well drawn as the first and second requirements will permit.

Correctness. This cannot be emphasized too much. A drawing that is not correct in every particular is not good for much and, in some cases, worse than useless, causing serious loss of time and material.

The writer recalls a case in point in which the drawing of one of a set of boiler plates contained a wrong dimension. The order was for thirty locomotives to be delivered on a certain date under penalty. The thirty wrong plates had to be thrown out and others ordered. The work was held up with serious loss of time, labor and material. The draftsman who made the blunder was discharged.

In addition to the correct mechanical drawing of the machine the drawing must contain all correct dimensions, notes, pattern numbers, finishes, etc.

In placing dimensions on a drawing, the draftsman should be able to put himself in the position of the workman who is to use it, and place the dimensions as far as possible where the workman would be most likely to look for them, this makes it easier to read the drawing and saves the workman's time.

There is always an inclination on the part of certain students in solving problems to copy the illustrations in the text, the finished drawings of other students or importune the Instructor to tell them just what to do without much effort or thought on their own part.

This is a great mistake on the students part, for, beyond the practice in mechanical drawing, he gets very little out of his machine-drawing course unless he realizes the design and construction of the machine he is drawing, why it is made so and not otherwise, how it is produced in the pattern shop, the foundry and finished in the machine shop, and put together on the erecting or assembling floor.

He should realize that the more he learns of the form, proportion and construction of the elements of machines contained in this course, the better foundation he will lay for his development as a designer of machines.

The young man who takes full advantage of this course, in Elementary Machine Drafting, will fit himself as a detail draftsman able to make commercial working drawings, and such ability will always be in demand at a good salary.

It is such work the young engineer is given to do first after graduation, and when he has been tested and his ability proved he is given more important and responsible work to do.

The principles of projection in the third angle (as given in Mechanical Drawing) will be used exclusively in making all drawings.

MECHANICAL DRAWING

CHAPTER I

INSTRUMENTS AND THEIR USES

1. **The Drawing Board** should be light for convenience in handling; material soft pine, constructed three-ply to prevent warping. The required size for this course is $16'' \times 21''$ $\times \frac{9}{16}''$.

FIG. 1.

Fig. 1 shows a drawing board that has given satisfaction in extensive use.

The left-hand edge should be true and square with the upper edge. It is not essential that the other edges should be perfectly square.

One face should be selected for the top face on which the drawing paper is to be pinned and when the left-hand edge has been made smooth and true it should be marked and always used for the T-square edge.

2. **The T-square** should be the same length as the board, viz., $21''$. There are many styles of T-squares made in different materials, but a well-made pearwood T-square with fixed head is comparatively low priced and quite suitable

for this work. The position at the drawing table when using the T-square is of some importance. As a rule the drafts-man should stand when pencilling in a drawing. It gives him more freedom in the use of his tools, saves time and is healthier than sitting crouched together on a stool.

When placing dimensions, lettering, notes, titles, bills of material or tracing, it is quite proper to sit on a stool of convenient height to give a good easy position when at work.

FIG. 2.

Most students, however, prefer to sit down when drawing at all times and some have good reasons for doing so, such as " tired " from standing so much in laboratory and shop work, " weak back," " feet hurt," etc.

A good easy position should be obtained when sitting down to draw with the light coming in from the left. See Figs. 2 and 3.

When drawing horizontal, straight lines the head of the T-square should be held rigidly against the left-hand edge of the drawing board, as shown in Fig. 3.

Horizontal straight lines should be drawn with the pencil held in a plane perpendicular to the board, passing through the edge of the T-square and making an angle of about

60° with the board. This angle should be maintained throughout the line, and drawn from left to right. See Fig. 4.

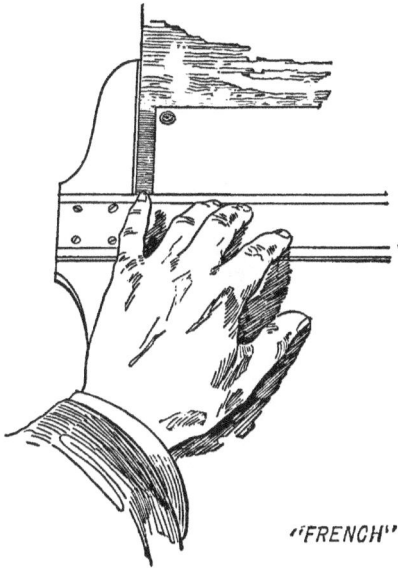

"FRENCH"

FIG. 3.

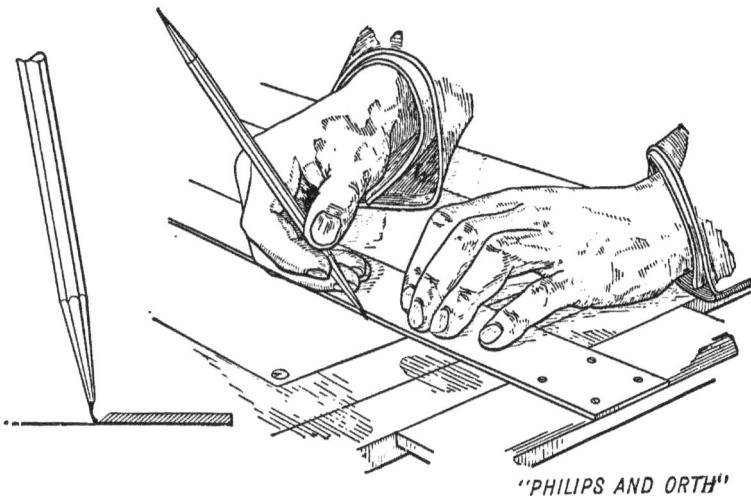

"PHILIPS AND ORTH"

FIG. 4.

3. **The Triangles,** one $30°\times60°\times10''$ long and $45°\times8''$ long, like those shown in Fig. 5.

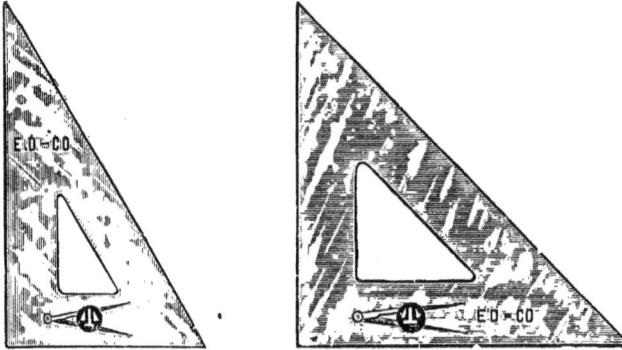

FIG. 5.

The material of the triangles should be celluloid; it is comparatively cleaner and more durable than wood or black rubber.

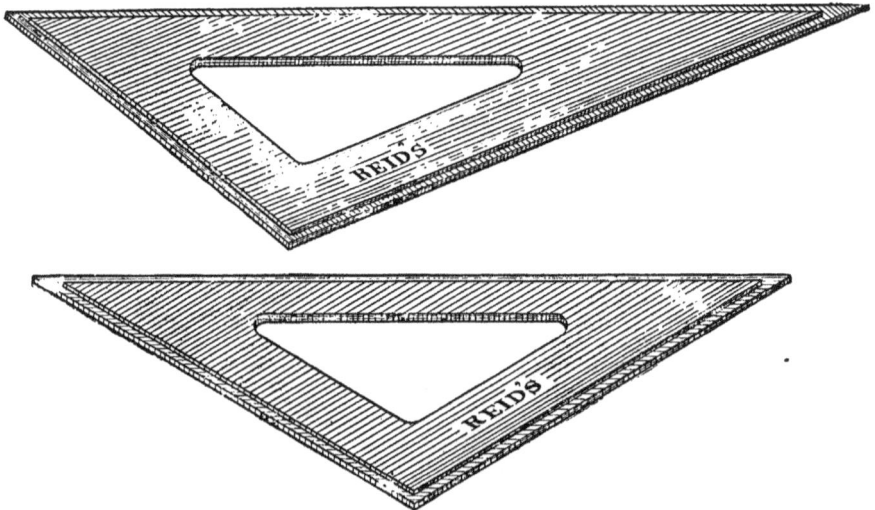

FIG. 5a.

The triangles shown at Fig. 5a have rabbeted edges; they were devised to raise the working edge from the paper to prevent blots when drawing ink lines to join other lines already inked. When drawing an ink line with the ordinary triangle, which is usually quite thin, great care must be observed when approaching another ink line at the end of the stroke, for unless the pen is quickly lifted from the paper as soon as it touches the line already inked, the ink in the pen seems to be drawn into the triangle, spreading under it and causing a bad smear. The raised edge prevents this.

All vertical lines should be drawn with the triangles and T-square, holding the pencil as explained in Art. 2. See Fig. 6.

FIG. 6.*

With the T-square, the $30° \times 60°$ and $45°$ triangles, various angles may be drawn as shown in Fig. 7.

At the right in Fig. 7, is shown how a hexagon may be drawn circumscribing a circle.

This is the method used in drawing hexagonal bolt-heads and nuts.

With the two triangles, parallel lines may be drawn to any given line as follows:

Place one edge of a triangle even with the given line, then place the long edge of the other triangle against another edge of the first triangle, and holding the second triangle firmly in position, slide the first triangle against the second

* From French's Engineering Drawing.

and any number of parallel lines may be drawn to the given line. See Fig. 8.

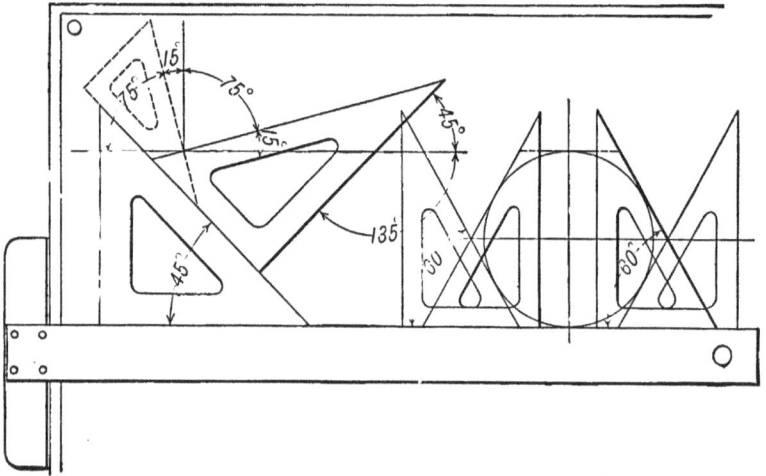

FIG. 7.

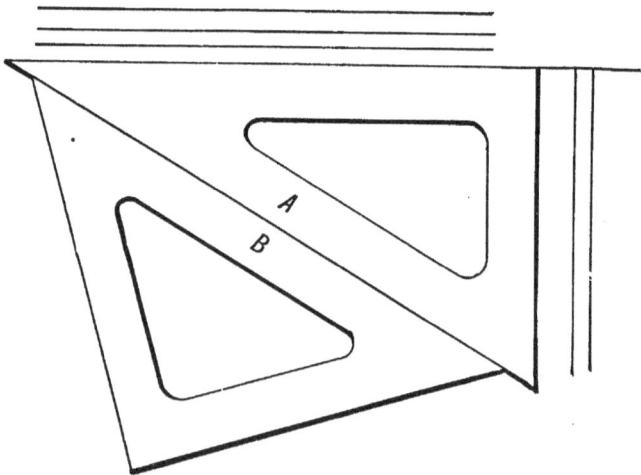

FIG. 8.

Lines perpendicular to given lines may be drawn with the triangles in a similar manner.

Parallel lines and perpendiculars may also be drawn with the T-square and one triangle.

4. The Pencil. Designs of all kinds are usually worked out in pencil first, and if to be finished and kept they are inked in and sometimes shaded; but if the drawing is only to be finished in pencil, then all the lines except construction, center, and dimension lines should be made broad and dark, so that the drawing will stand out clear and distinct. It will be noticed that this calls for two kinds of pencil-lines,

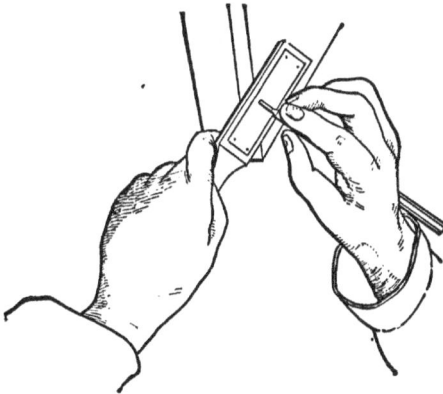

FIG. 9.

the first a thin, even line made with a hard, fine-grained lead pencil, not less than 6H (either Koh-i-noor or "Eldorado"), and sharpened to a knife-edge in the following manner: The lead should be carefully bared of the wood with a knife for about $\frac{1}{2}''$, and the wood neatly tapered back from that point about $\frac{1}{4}''$; then lay the lead upon the emery-paper sharpener illustrated in Fig. 9 and carefully rub to and fro until the pencil assumes a long taper from the wood to the point; now turn it over and do the same with the other side, using toward the last a slightly oscillating motion on both sides until the point has assumed a sharp, thin, knife-edge endwise and an elliptical contour the other way.

This point should then be polished on a piece of scrap drawing-paper until the rough burr left by the emery-paper is removed, leaving a smooth, keen, ideal pencil-point for drawing straight lines. See *A* and *B*, Fig. 10.

With such a point but little pressure is required in the hands of the draftsman to draw the most desirable line, one that can be easily erased when necessary and inked in to much better advantage than if the line had been made with a blunt point, because, when the pencil-point is blunt the

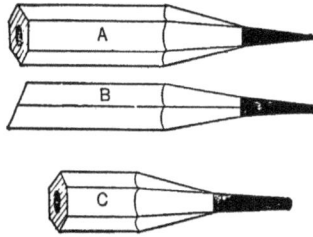

FIG. 10.

inclination is to press hard upon it when drawing a line. This forms a groove in the paper which makes it very difficult to draw an even inked line.

The second kind of a pencil-line is the broad line, as explained above; it should be drawn with a somewhat softer pencil, say 4H, and a conical point.

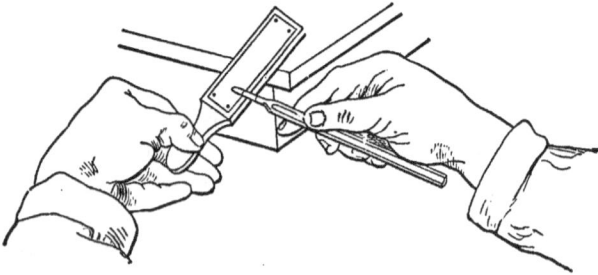

FIG. 11.

The operations for sharpening the 4H pencil are the same as for the 6H, except that instead of rubbing flat, the pencil should be rotated in the fingers while it is being rubbed to and fro on the pointer (see Fig. 11), pressing slightly toward the point so as to form a conical-shaped point. The point should not be sharp, like a needle, but round and sharp enough to give a clear, dark, strong line.

5. Pocket Case of Drawing Instruments. It is a common belief among students that any kind of cheap instrument will do with which to learn mechanical drawing, and not until they have acquired the proper use of the instruments should they spend money in buying a first-class set. This is one of the greatest mistakes that can be made. Many a student has been discouraged and disgusted because, try as he would, he could not make a good drawing, using a set of instruments with which it would be difficult for even an experienced draftsman to make a creditable showing.

If it is necessary to economize in this direction it is better and easier to get along with a fewer number, and have them of the best, than it is to have an elaborate outfit of questionable quality.

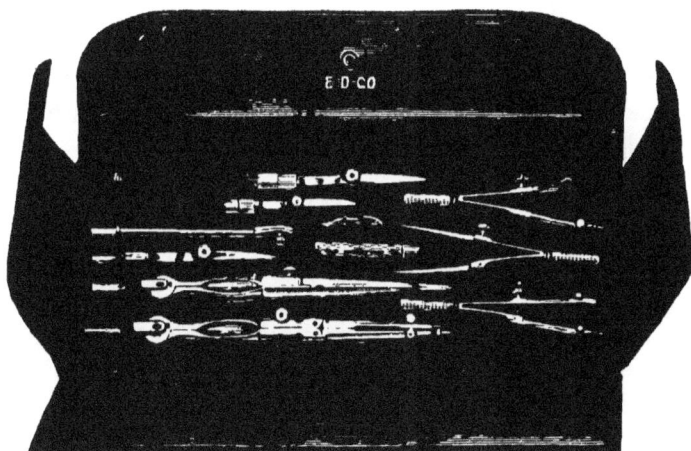

FIG. 12.

The instruments shown in Fig. 12 are well made, of a moderate price, and with care and attention will give good satisfaction for a long time.

This set consists of:

1 large compass, with pen and pencil points and lengthening bar.

1 large divider, sometimes called spacer.

3 spring bow instruments for pen, pencil and spacer.

2 straight-line drawing pens, medium and short size.
1 pencil point holder.
1 compass joint tightener.

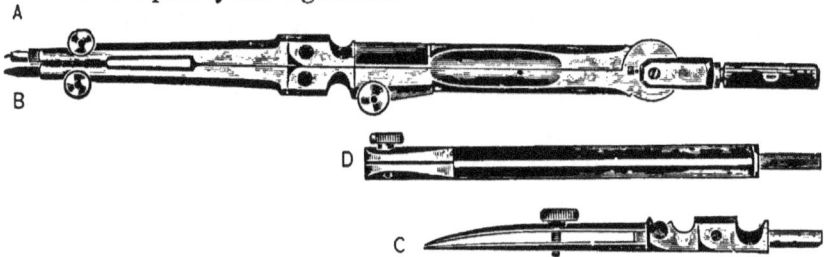

FIG. 13.

6. The large compass in detail is shown in Fig. 13. *A* is the needle leg; *B*, the pencil leg; *C* the pen, and *D*, the lengthening bar.

This instrument is used for drawing arcs and circles larger than can be properly drawn with the spring bows (Fig. 19).

Circles of about $3\frac{1}{2}''$ diameter may be drawn with the legs straight. See Figs. 15 and 16.

FIG. 14.

The method of operation when using the large compass is as follows: Adjust the needle point so that its flat side is next to the pencil or pen and its point about $\frac{1}{32}''$ longer than the point of the pencil or pen. When the point, through which the curve is to be drawn, has been accurately marked and the center located, guide the needle point to the center

with the little finger of the left hand, see Fig. 14, and draw
the curve exactly through the mark already located with a
clockwise motion inclining the instrument a little toward the
direction of the line, see Figs. 15 and 16, which illustrate the
beginning and ending of the motion.

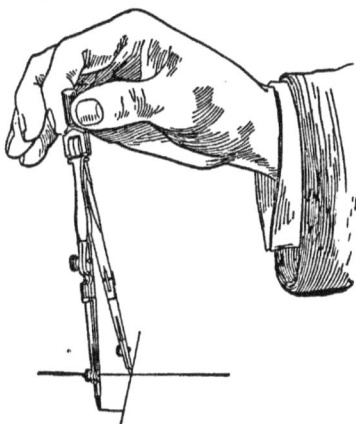 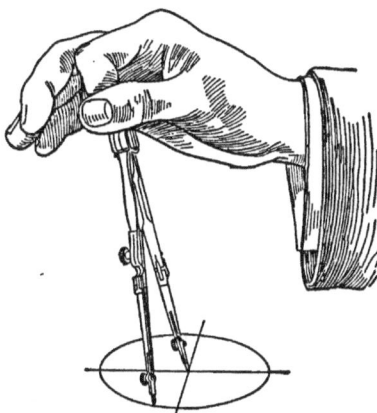

FIG. 15.   FIG. 16.

When circles of about 10″ diameter are to be drawn the
leg of the compass should be bent at the knuckle joints so
that the pencil or pen leg and the needle leg will both be

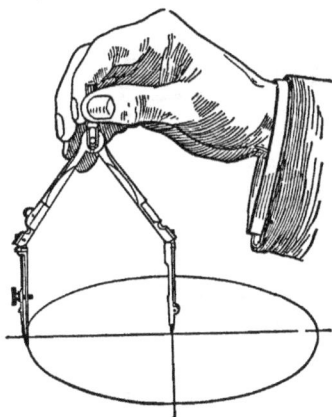

FIG. 17.

perpendicular to the paper to provide a sharp even line
throughout its length, Fig. 17.

Circles larger than 10″ diameter up to about 14″ may be drawn with the use of the lengthening bar shown at *D* in Fig. 13.

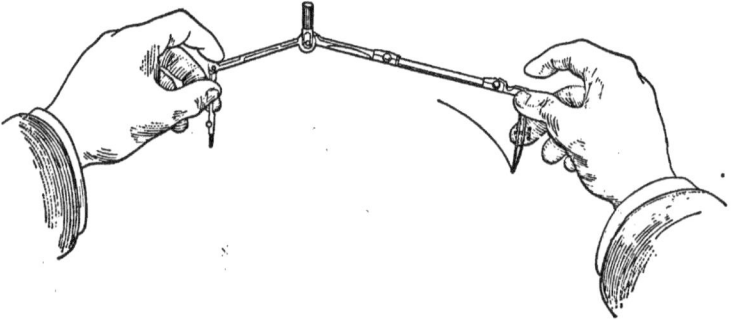

FIG. 18.

To use the lengthening bar, withdraw the pencil or pen leg, insert the bar, add the pencil or pen leg and bend it and the needle leg at the knuckle joint and the instrument will be ready for use as shown in Fig. 18.

7. **The Spring Bow Instruments,** pen, pencil and spacer, Fig. 19. The pencil and pen bows are for very small arcs and circles, such as joining straight lines with fillets, etc.

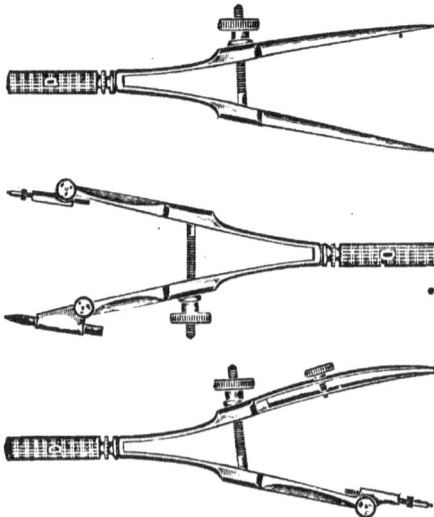

FIG. 19.

These are very important instruments, because the beginner by their use is able to make a uniformly good drawing. While the experienced draftsman may put in small fillets and round corners free-hand, he can take this liberty because he has already learned the use of the spring bows and also because he has learned to do such freehand work well, but the beginner has no such experience and should, therefore, practice the use of the bow instruments at every opportunity.

Before inking or tracing a drawing, all small arcs, such as fillets and round corners, and small circles should be carefully pencilled in with the bow instruments; much better work is obtained than is probable by the beginner who, thinking to save time and effort, tries to ink these small curves without pencilling. Many otherwise good drawings are spoiled in appearance because of the bad joints between curves and straight lines.

Small arcs, circles and all curved lines of any description should be inked in all over the drawing before any straight lines are inked; this is essential to obtain the best results in tracing a drawing.

It is much easier to know where to stop the arc line, and to draw the straight line tangent to it than it is to reverse the process.

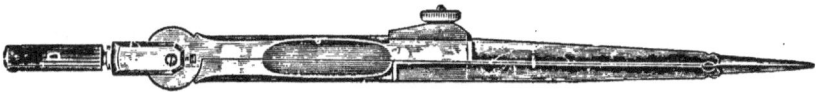

FIG. 20.

8. The Large Dividers or Spacers, Fig. 20. This instrument should be held in the same manner as described for the compass. It is very useful in laying off equal distances on straight lines or circles. To divide a given line into any number of equal parts with the dividers, say 12, it is best to divide the line into three or four parts first, say 4, and then when one of these parts has been subdivided accurately into three equal parts, it will be a simple matter to

step off these latter divisions on the remaining three-fourths of the given line. Care should be taken not to make holes in the paper with the spacers, as it is difficult to ink over them without blotting.

To Divide the Line by Measurement. First ascertain the length of the given line by measuring with rule or scale. Let the length be assumed to be 12″ and the line is to be divided into 12 equal parts as before. Such division in this case will equal exactly 1″. Set the dividers to 1″ on the rule or scale and step off the 12 equal parts on the given line, adjusting the dividers until the line is divided exactly into the required number of equal parts.

9. The Straight-line Pen, Fig. 21. There are usually two pens in a case of instruments, one about 5″ long and a smaller one. The small one is not of much use. The theory is that the small pen is for drawing fine lines but in fact the larger pen will give just as fine lines and is easier to handle.

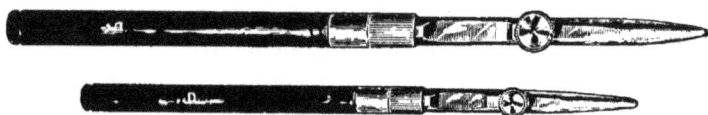

FIG. 21.

The best form for a straight-line pen, in the writer's opinion, all things considered, is that shown in Fig. 21. The spring on the upper blade spreads the blades sufficiently apart to allow for thorough cleaning and sharpening.

The pen should be held in a plane passing through the edge of the T-square at right angles to the plane of the paper, and making an angle with the plane of the paper ranging from 60° to 90°. See Fig. 22.

The blades of the pen should be of equal length and when held against the T-square or triangle with the blades parallel to their edge, the pen is guided by the upper edge of the T-square or triangle and the point of the pen is held away from the lower edge as illustrated at *B*, in Fig. 23.

If the pen should be held out of the perpendicular, as

shown at *C*, the result would probably be a ragged, uneven line; if held as shown at *A*, there is danger of the ink running under the blade of the straight edge and causing a blot.

FRENCH

FIG. 22.

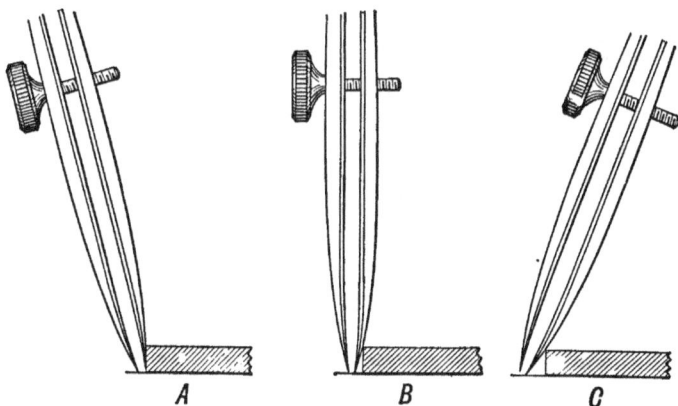

A   B   C

FIG. 23.

10. Sharpening the Pens. The best of drawing-pens will in time wear dull on the point, and until the student has learned from a competent teacher how to sharpen his pens it would be better to have them sharpened by the manufacturer.

It is difficult to explain the method of sharpening a drawing-pen.

If one blade has worn shorter than the other, the blades should be brought together by means of the thumb-screw, and placing the pen in an upright position draw the point to and fro on the oil-stone in a plane perpendicular to it, raising and lowering the handle of the pen at the same time, to give the proper curve to the point. The Arkansas oil-stones are best for this purpose.

The blades should next be opened slightly, and holding the pen in the right hand in a nearly horizontal position, place the lower blade on the stone and move it quickly to and fro, slightly turning the pen with the fingers and elevating the handle a little at the end of each stroke. Having ground the lower blade a little, turn the pen completely over and grind the upper blade in a similar manner for about the same length of time; then clean the blades and examine the extreme points, and if there are still bright spots to be seen continue the grinding until they entirely disappear, and finish the sharpening by polishing on a piece of smooth leather.

The blades should not be too sharp, or they will cut the paper. The grinding should be continued only as long as the bright spots show on the points of the blades.

11. The Triangular Scale, Fig. 24. This scale, illustrated in Fig. 24, was arranged to suit the needs of the students in machine drawing. It is triangular and made of boxwood. The six edges are graduated as follows: $\frac{1}{16}''$ or full size, $\frac{1}{30}''$, $\frac{3}{4}''$ and $\frac{3}{8}'' = 1$ ft., $1''$ and $\frac{1}{2}'' = 1$ ft., $3''$ and $1\frac{1}{2}'' = 1$ ft., and $4''$ and $2'' = 1$ ft.

Drawings of very small objects are generally shown enlarged—e.g., if it is determined to make a drawing twice the full size of an object, then where the object measures $1''$ the drawing would be made $2''$, etc.

Larger objects or small machine parts are often drawn full size—i.e., the same size as the object really is and the drawing is said to be made to the scale of full size, or $12'' = 1$ ft.

Large machines and large details are usually made to a reduced scale—e.g., if a drawing is to be made to the scale of $2'' = 1$ ft., then $2''$ measured by the standard rule would be

divided into 12 equal parts and each part would represent
1″. See Fig. 24.

When laying off a dimension on a drawing with scale

FIG. 24.

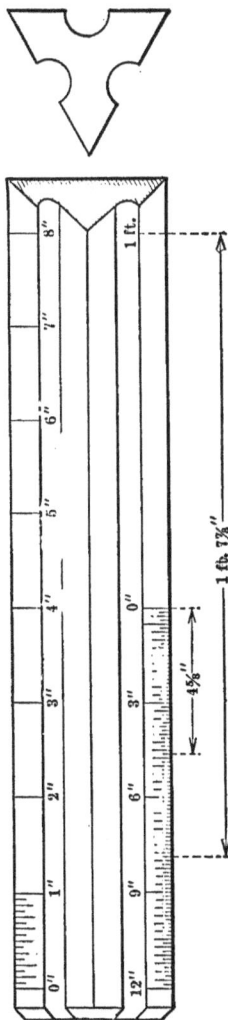

FIG. 25.

it is bad form to use the compass and dig the needle into
the scale and measure the dimension with the instrument on
the scale; it hurts the compass and mars the divisions on
the scale. The best way is to lay the scale on the drawing

and with a sharp-pointed pencil mark the distance directly
on the drawing. See Fig. 26.

FIG. 26.

To lay off feet and inches, see Fig. 25.

12. The Protractor, Fig. 27. This instrument is for
measuring and constructing angles. It is shown in Fig. 27.
It is used as follows when measuring an angle: Place the
lower straight edge on the straight line which forms one of
the sides of the angle, with the nick exactly on the point of
the angle to be measured. Then the number of degrees
contained in the angle may be read from the left, clockwise.

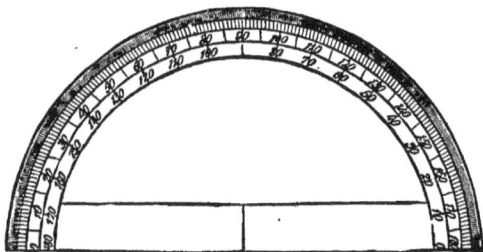

FIG. 27.

In constructing an angle, place the nick at the point from
which it is desired to draw the angle, and on the outer circumference of the protractor, find the figure corresponding

to the number of degrees in the required angle, and mark a point on the paper as close as possible to the figure on the protractor; after removing the protractor, draw a line through this point to the nick, which will give the required angle.

13. **The Irregular Curve** is made of celluloid and should be of a form suitable to meet most needs in drawing smooth irregular lines. The Curve shown in Fig. 28 is useful for

FIG. 28.

drawing irregular curves through points that have already been found by construction, such as ellipses, cycloids, epicycloids, etc., as in the cases of gear-teeth, cam outlines, rotary pump wheels, etc.

When using these curves, that curve should be selected that will coincide with the greatest number of points on the line required.

The Curve should coincide with not less than three points at each drawing.

14. **The Emery Pencil Pointer**, Fig. 29. The lead or graphite in the pencil must not be cut with the knife. After the wood has been removed the lead should be sharpened

FIG. 29.

on the Emery Pencil Pointer in the manner described in Section 4 and Figs. 9, 10, and 11. The pointer shown in Fig. 29 is the best style for handling and saves the soiling of the fingers.

15. **The Soft Rubber Eraser**, Fig. 30. When erasing pencil lines, the rubber should not be pushed hard into the

paper, as the inclination is, but hold the eraser lightly in the fingers and try to erase the line without hurting the paper.

When removing ink lines from tracings it used to be thought necessary to use a steel or glass and rubber eraser, but the writer found this to be unnecessary. It takes an expert to use a steel eraser without damaging the paper and the glass rubber will always leave an ugly mark which will show later in the blue print.

FIG. 30.

To erase ink lines or blots from tracings, be sure the ink is perfectly dry, then place a hard, smooth surface like a triangle under the ink to be erased and holding the tracing cloth tightly over it with the fingers of the left hand use the soft rubber with short, quick strokes and rub off the ink, not *in*. It will take a little longer to make the erasure complete than it would using the steel or glass rubber but it gives much better results.

When pencil or ink lines are to be erased be sure to make a thorough job of it. If the erasing is not complete it will always have a bad appearance.

16. An erasing shield is sometimes used to take out short lines and spots. The idea is to save rubbing the adjacent lines and confine the rubbing to the ink to be erased. It is the opinion of the writer, however, that an erasing shield is unnecessary, in fact may be a detriment to obtaining good results. When using the shield the rubbing is confined to a small area and the inclination is to press hard while rubbing with consequent danger of rubbing through the paper. .

17. The "Art-gum," Fig. 31, is a soft material used for cleaning the drawing after all the inking is done. It takes the place of and is better than sponge rubber.

Most of the cleaning should be done before inking. Gasoline is good for cleaning tracings.

Black India Ink, Fig. 32, is a liquid waterproof Chinese ink. It should be fresh to obtain the best results.

If the ink has become thickened in hot weather it may be thinned by adding a few drops of water. Old, stale, gritty ink will give trouble.

Thumb Tacks for fastening the drawing paper to the board are made of stamped steel. Heads of $\frac{3}{8}''$ diameter are large enough. Lines should not be drawn with the T-square resting on thumb tacks. Remove the tacks temporarily so that the T-square may lie flat on the paper.

FIG. 31.   FIG. 32.   FIG. 33.

Thumb tacks come one dozen in a small box like that shown in Fig. 33.

18. Cross-section Pad. This pad is $8'' \times 10''$ in size and is divided 8×8 to the inch. It is useful in making the eight sheets of lettering included in this course.

It is also used in machine sketching, taking notes, figuring calculations in determining machine proportions, etc.

Drawing Paper. Use cream detail paper $15'' \times 20''$. This paper is cut exactly to size so there will be no margin to cut off. A border line is to be drawn as follows: $1\frac{1}{2}''$ from left-hand edge of paper and $\frac{1}{2}''$ from all the other edges.

Tracing Cloth. This should be of the grade called "Imperial." Always use the dull side of the tracing cloth. See the answers to Question 31 in "Present Practice in Drafting Room Conventions," in Appendix, page 225.

CHAPTER II

STANDARD DRAFTING ROOM CONVENTIONS

19. In most drafting rooms, each draftsman is supplied with a set of conventions, rules of procedure in making working drawings, etc.

Endeavors have been made, from time to time, with a large measure of success, to establish uniform conventions, rules and methods in making commercial drawings.

The importance of a uniform system is apparent when we realize that shopmen have often to use drawings made by the draftsmen of different companies.

The following rules and conventions adopted for this work have been found by investigation to be of nearly universal practice in all the leading and progressive drafting rooms in the United States.

20. Pencil Drawings. Unless otherwise ordered all pencil drawings shall be made on cream detail paper with a 6H pencil sharpened as directed in Section 4. This applies to both straight-line and compass pencil.

Pencil drawings are to be finished and all lettering, figures and arrow points made with the 4H pencil, sharpened to a conical point.

On all pencil drawings the *title, bill of material* and *border lines* are to be inked.

21. Tracings are to be made on the dull side (see answer to Question 31, page 225 in Appendix) of " Imperial " tracing cloth. When erasing is necessary use " Emerald " rubber and triangle.

Soiled tracings can be cleaned with gasoline or benzine.

22. Lettering. The lettering on all drawings shall be freehand, sloping, Gothic capitals, of uniform height. (See answer to Question 8 in Appendix.) All notes on drawings are to be $\frac{3}{32}''$ high. The size of figures in dimensions may vary according to conditions. The dividing line in fractions must always be made horizontal.

23. Bill of Material. When a bill of material is given on an Elementary Machine Drawing it should conform in every respect to that shown in Fig. 34.

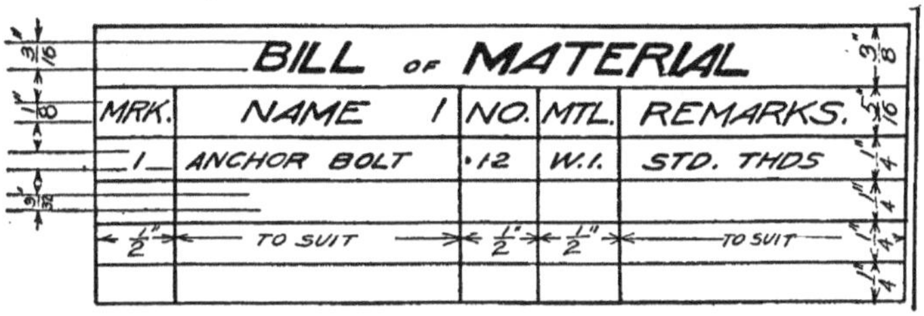

FIG. 34.

The letters in the words " Bill of Material " are to be inked with heavy lines as shown.

The " Ball " point pen No. 516 is to be used for all letters and figures.

The extra width of lines in letters that are to be made *heavy* should be applied with the " Gillott " pen No. 303 to insure a sharp even outline.

The lowest line in the table of the bill of material should be drawn $\frac{1}{4}''$ above the highest line of the title.

The outer boundary lines and the line under the words " Bill of Material " are to be inked with heavy lines as shown in Fig. 34.

24. Standard Title. All drawings in mechanical and machine drawing shall have a title conforming to the standard title shown in Fig. 35.

The title will occupy a space of about $2'' \times 4\frac{1}{2}''$ and be placed at the lower right-hand corner of the plate $\frac{1}{16}''$ inside of the border line.

There shall be no boundary line drawn around the title.

The guide lines for the lettering in the title and bill of

material should be drawn very narrow and light and
erased as far as possible before inking the lettering.

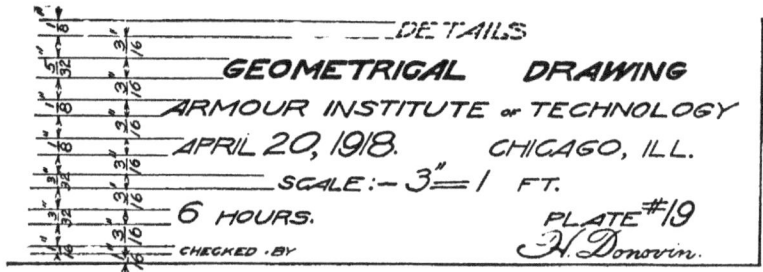

FIG. 35.

25. Standard Lines.

Visible Object Lines

Visible object lines in ink on tracings shall be not less
than $\frac{1}{50}''$, nor more than $\frac{1}{32}''$, wide.

A ———————————————— Visible object line $\frac{1}{50}''$

Invisible Object Lines

B ═══ ═══ ═══ ═══ ═══ ═══ Invisible object lines $\frac{1}{75}''$

Invisible object lines in ink shall be about $\frac{1}{75}''$ wide. The
dashes are to be not less than $\frac{1}{8}''$ nor more than $\frac{3}{16}''$ in length,
$\frac{1}{8}''$ for short dash lines and $\frac{3}{16}''$ for long dash lines.

The spaces between dash lines are to be very short, just
enough to show that the line is broken.

When two invisible object lines have to be drawn close
together, as shown at B, the breaks should be evenly arranged under one another. It is a much neater arrangement
than that shown at C.

C ══ ══ ══ ══ ══ ══ ══ ══ Incorrect arrangement

Center Line

Center lines shall have a long dash and a short one
alternately. In a long center line the long dash may be 3″
long and the short dash ⅛″ and the spaces between them
1/32″. In short center lines the long dash may be made to
suit but the short dash should always be about ⅛″ long.
See *D*.

D ——— — ——— — ——— —— ——— ——— · Center line

The center line must be made very narrow. If the pen
is worn so that it will not make a narrow line stop using it
until it is sharpened.

Dimension Line

E ◄——————   ——————► Dimension line

Dimension lines, as shown at *E*, are to be very narrow
with a suitable break for the dimension and usually an arrow
point at each end.

Arrow points should be sharp at point and narrow at the
wings. Radial dimension lines drawn from a center should
have no arrow point at the center.

Witness Line

━┳━ ——— ——— —— ——— Witness line

Witness lines are short lines extending from any part of
a drawing to meet the dimension line outside of the part
to be dimensioned. Witness lines must be very narrow; it is
a broken line, one short dash about ⅛″ long, one break and
another longer dash to pass the arrow point on the dimension line about 1/16″. The dimension line should be far
enough away from the part to be dimensioned to give plenty
of room for the dimension figures without crowding.

Dimension figures should not be placed close to lines or other figures. Dimensions may be placed upon the drawing, sometimes with good effect, but a good general rule is to place all dimensions outside of a piece whenever it is convenient to do so.

Limiting Break Line

———————————————————————— Limiting break line

When only a small portion of an object needs to be shown it should be limited by drawing a break line, as above, freehand, with a lettering pen. The line should not be very ragged, just waved enough to show the character of the line.

Adjacent Part Line

— — — — — — — — — — Adjacent part line

Adjacent part lines are sometimes shown to indicate a part connected with that for which the drawing is made but is not an essential part of that drawing.

The line is made with $\frac{1}{4}''$ dashes of medium weight and short spaces. See Fig. 115, which illustrates the use of an adjacent part line.

Alternate Position Line

·——·——·——·——·——·—— Alternate position line

An alternate position line is used to represent a limit position of a moving part. The base outline of the object may be shown at one or more of its extreme positions, while the regular drawing shows it at its central position. The line is made of fine dashes $\frac{3}{4}''$ and $\frac{1}{8}''$ long, alternately. The space between the dashes should be quite short. The alternate positions of the crosshead on an engine is an example.

Cutting Plane Line

——— — — ——— — — ——— — — ——————— Cutting plane line

A cutting plane line is used to indicate on a drawing where a sectional view is to be taken other than on a center line. The character of the line, as shown above, is composed of dashes $\frac{3}{4}''$ long divided by two short dashes about $\frac{1}{16}''$ long each. The spaces between all the dashes to be very short. The dashes are to be of medium width, say about $\frac{1}{50}''$.

BORDER LINES:

REFERENCE ARROW
        LINES

Should always be drawn straight with ruling pen and set obliquely, i.e., neither vertically nor horizontally.

Very heavy $\frac{1}{16}''$ wide

Hatch Lines

Hatch lines are used to indicate parts of a drawing in section. They are sometimes called section lines. They should be drawn $\frac{1}{16}''$ apart at an angle of 45°.

When two adjacent surfaces in section are to be hatch-lined the hatch lines on the two surfaces should be drawn in opposite directions. When three or more surfaces of different objects come together in a section drawing the hatch lines must be drawn at two different angles and two opposite directions.

26. **Standard Hatch Lines.** Conventional hatch lines are placed on drawings to distinguish the different kinds of materials used when such drawings are to be finished in pencil or traced for blue printing, or to be used for a reproduction of any kind.

Fig. 36 shows a collection of hatch-lined sections that is now the almost universal practice among draftsmen in this and other countries, and may be considered standard.

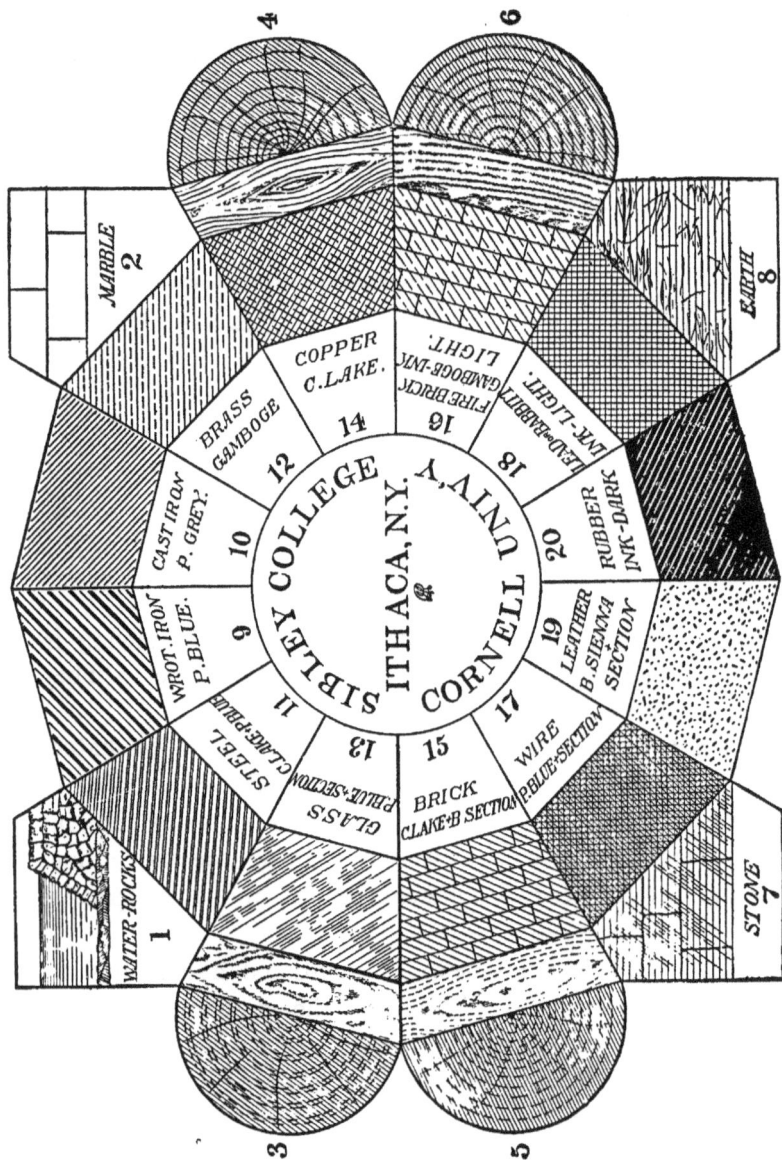

FIG. 36.

Conventional Breaks

27. Conventional Breaks. Breaks are used in drawings sometimes to indicate that the thing is actually longer than it is drawn, sometimes to show the shape of the cross-section and the kind of material. Those given in Fig. 37 show the usual practice.

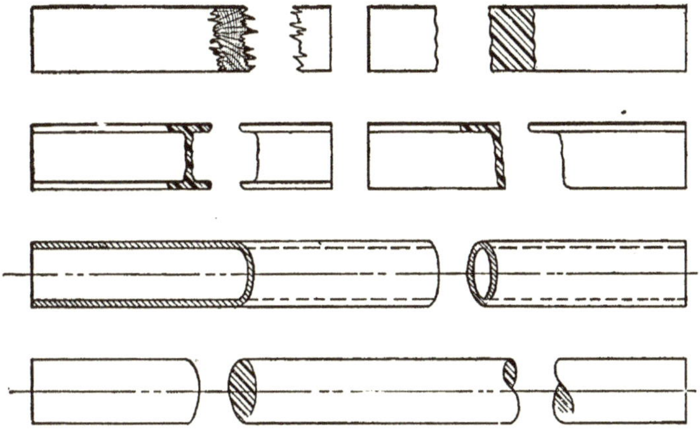

FIG. 37.

28. Finishes.

Finish. f.

Shape to dimensions with a cutting tool: The surface to conform to good shop practice. Finish for appearance sake only when so specified.

Trim

Shape by any convenient method such as chipping, filing, grinding, etc.

Spot Face

Use counterbore or similar tool to make the spot true.

Polish

Make the surface smooth and glossy when used without connection with the word " finish " or the finish mark " f ";

neither a perfectly true surface nor measured adherence to dimensions is called for. A suitable polish may be had by first grinding and then buffing.

Finish and Polish

Shape to dimensions with a cutting tool and make the surface smooth and glossy without destroying the accuracy of the work.

Grain

Give the surface the appearance of having a straight grain, by rubbing with emery cloth in a direction parallel with one of the edges of the surface.

Finish and Grain

Shape to dimensions with a cutting tool and afterwards increase the trueness of the surface by scraping.

Finish and Grind

Shape to dimensions with a cutting tool and afterwards increase the trueness of the surface by grinding.

Matt

Cover uniformly with indentations, touching one another and all of the same size, shape and depth.

Blue

Heat uniformly until the color changes to blue.

Nickel Plate

Electroplate with nickel.

Lacquer Blue

Coat with blue lacquer.

Dip and Lacquer

Cleanse by immersing in acid, afterwards coat with yellow lacquer.

Boil in Oil

Immerse in boiling linseed oil.

Burn in Oil

Heat to dull red and plunge into linseed oil.   Or dip cold into linseed oil and afterwards burn the oil off.

French Polish

Coat with shellac and boiled linseed oil by the process known as French polishing.

Shellac

Coat with a suitable thickness of shellac.

Black Japan

Bake upon the surface a suitable thickness of black enamel paint.

Dip in Linseed Oil

Immerse in boiled linseed oil.

Paraffin

Immerse in boiling paraffin.

Sand Rub and Oil

Rub with sandstone, wipe clean and apply boiled linseed oil.

Paint

Coat with a suitable thickness of paint.

Knurl

Using a knurling tool make the surface suitable for gripping. There are two standard styles of knurling.

1. Straight knurl for round edge grips *A* not to exceed $\frac{1}{8}''$.

2. Diamond knurl, standard medium.

Pene

Close the fibers of the material or produce closer contact between pieces by hammering.

Designate finishes on drawings as follows:

Finish

In general indicate by applying *f* on the edge line of the surface to be finished thus: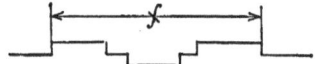
When a surface is of such shape as to require several finish marks inconveniently close together the finish may be indicated thus: This means that the entire surface between the arrow points is to be finished. If desired an explanatory note may be used as " Finish all over " using the complete word finish.

Trim

Indicate by an explanatory note as " trim flush with edge of flange " using the complete word trim.

Spot Face

Indicate thus: " Spot face."

Polish

Indicate by note, as " Polish exposed surfaces," or thus:
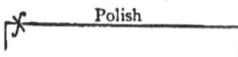 When desirable indicate thus: " Finish and Polish from *A* to *B*."

Finish Smooth

In general indicate thus:
or thus: Finish smooth from
A to *B*.

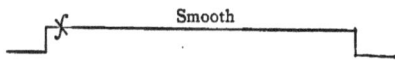

Finish and Scrape

Thus:

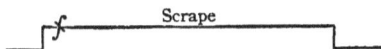

Finish and Grind

Thus:

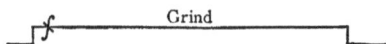

Finish and Grain

Thus:

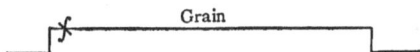

or " Face and grain exposed surfaces."

Matt

Thus:

Nickel Plate

Indicate by note using abbreviation " N.P."

All other finishes

As Blue, Copper Plate, Copper Plate and oxidize, Lacquer, Boil in Oil, French Polish, Shellac, Black Japan, Dip in Linseed Oil, Paraffin, Sand Rub and Oil, and Paint indicate by note in Gothic letters $\frac{3}{32}''$ high.

29. Abbreviations. Abbreviations are useful in saving time and space. Most of the abbreviations given here are common and well known. The meaning of uncommon abbreviations is suggested by preceding words or by the drawing.

Alternating current....................A.C.
Amperes...............................Amp.
Aluminum..............................Almn.

Babbitt.............................Bb.
Back gear..........................BGr.
Bevel gear.........................Bev.Gr.
Birmingham wire gauge..............B.W.G.
Board..............................Bd.
Bolt...............................B.
Bracket............................Bkt.
Brass..............................Br.
Bronze.............................Brz.
Building...........................Bldg.
Button head bolt...................Btn.Hd.B.
Cabinet............................Cab.
Candle power.......................C.P.
Cap screw..........................Cap.Sc.
Carriage...........................Car.
Case harden........................C.H.
Cast brass.........................C.B.
Cast copper........................C.Cop.
Cast iron..........................C.I.
Cast steel.........................C.S.
Center line........................C.
Center.............................Ctr.
Change gear........................Ch.Gr.
Circular pitch.....................C.P.
Cold-rolled steel..................C.R.S.
Company............................Co.
Compound rest......................Comp.R.
Conical............................Λ
Counterbore........................C.br.
Countersink........................Csk.
Counter shaft......................Co.Sh.
Centimeter.........................Cm.
Circumference......................Circum.
Corona steel.......................Cro.S.
Crank..............................Cr.
Crucible steel.....................Cru.S.
Cylinder...........................Cyl.

Degrees Centigrade...................15° C.
Degrees Fahrenheit...................15° F.
Department..........................Dept.
Diameter............................Dia.
Diametral pitch.....................D.P.
Direct current......................D.C.
Double back gear....................D.B.Gr.
Double bevel gear...................D.Bev.Gr.
Double chamfered hexagon nut.........Dbl.Chmfd.Hex.Nut.
Drawing............................Dwg.
Drill..............................Dl.
Electrical.........................Elect.
Electromotive force.................E.M.F.
Experiment.........................Exp.
Eye bolt...........................EyeB.
Feather............................Fthr.
Feet or foot.......................Ft.
Feet or foot and inches.............2'4".
Fillister head brass machine screw......Fil.Hd.B.M.Sc.
Fillister head iron machine screw.......Fil.Hd.I.M.Sc.
Finish.............................f.
Flat head stove bolt.................Fl.H.StoveB.
Flat head wood screw................Fl.H.Wd.Sc.
Flexible...........................Flex.
Force or weight....................F.f.
General............................Gnl.
Half nut...........................H.N.
Hand wheel.........................H.W.
Hardened...........................H.
Head...............................Hd.
Head stock.........................Hd.St.
Headless set screw..................Hdlss.Set Sc.
Hexagon............................Hex.
Hexagon nut........................Hex.Nut.
Horsepower.........................H.P.
Inch or inches.....................In., Ins. or "
Kilowatts..........................K.W.

Kilogram	Kg.
Kilometer	Km.
Lag screw	Lag Sc.
Lead	Lead.
Lead screw	L.Sc.
Length	L. or l.
Long	Lg.
Lower side	L.Sl.
Machine	Mach.
Machine screw	M.Sc.
Machinery steel	M.S.
Malleable iron	Mal.I.
Manufacturing	Mfg.
Material	Mtl.
Maximum	Max.
Medium steel	Med.S.
Metres	M.
Mild steel	St.
Milled body tap bolt	M.B.Tap B.
Millimeters	mm.
Minimum	Min.
Moment of inertia	K.
Negative	Neg.
Not to scale	N.T.S.
Nickel plate	N.P.
Number for designating	No. or ⌗.
Number: quantity	Nbr.
Octagonal	Oct.
Open-hearth steel	O.H.S.
Oval	O.
Ounce or ounces	Oz.
Pattern number	Patt. ⌗.
Per cent	%
Pieces	Pcs.
Pitch	Pi.
Pitch diameter	P.D.
Phosphor	Ph.

Phosphor brcnze.......................Ph.Brz.
Polish...............................Pol.
Positive.............................Pos.
Power...............................P.
Power feed...........................P.F.
Pressure.............................p.
Quadrant............................Quad.
Radius...............................R. or Rad.
Railway..............................Ry.
Ream................................Rm.
Revolutions per minute................R.P.M.
Rough...............................R.
Screw...............................Sc.
Seamless............................Smlss.
Set screw............................Set Sc.
Shaft................................Sh.
Shoulder screw.......................Sh.Sc.
Sketch...............................Sk.
Specification.........................Spec.
Square...............................Sq.
Square feet..........................□′
Square inch..........................□″
Standard.............................Std.
Stationary part......................Stator.
Steel................................S.
Steel castings........................S.C.
Stud bolt............................Stud B.
T-head bolt..........................T.Hd.B.
Tap.................................Tp.
Temperature.........................Temp.
Threads.............................Thds.
Tool steel............................T.S.
Weight..............................Wgt.
Wrought iron........................W.I.

30. **Geometrical Definitions.** To those who have studied geometry the following definitions will simply be reminders of things learned and partially forgotten.

There are, however, some who have never studied geometry who believe that mechanical drawing would be a great assistance to them in their work.　They are right, there is no subject of such a practical nature known to the writer that can be so easily acquired and at such little cost of time, labor and money as mechanical drawing and be of such a benefit to them, no matter what their business may be in this age of practical engineering. A practical knowledge of mechanical drawing can be acquired without a previous study of geometry.　The following definitions will be illuminating and interesting to those taking up mechanical drawing before having studied geometry.

1. *Geometry* is the science of space, whether linear, superficial or solid.　It treats of points, lines, surfaces and solids, their construction and measurement.

2. A *Point* has position but neither length, breadth nor thickness.

3. A *Line* has length but no breadth or thickness.

4. A *Straight Line* or *Right Line* is the shortest distance between two points.

5. A *Curved Line* is a line no part of which is straight.

6. *Parallel Lines* are equidistant from each other in the same plane and never meet however far they may be produced.

7. A *Horizontal Line* is a line parallel to the horizon.　A horizontal line in a drawing is usually drawn with the T-square from left to right.

8. A *Vertical Line* is a line that is perpendicular to the plane of the horizon.　A still plumb line is a vertical.

In a drawing a vertical may be drawn up and down the paper with T-square and right-angled triangle.

9. *Oblique Lines* are neither horizontal nor vertical but are inclined to both.

10. *Perpendicular Lines*.　A line is perpendicular to another line when the angles on either side of it form two right angles.

Vertical and horizontal lines are always perpendicular to each other but all perpendicular lines are not horizontal and vertical.

11. *A Plane* is a surface such that if any two of its points be joined by a straight line that line will lie wholly in that surface. A surface has length and breadth without thickness.

12. A *Plane Figure* is a portion of a plane bounded on all sides by straight or curved lines.

13. A *Plane Figure* bounded by straight lines is a rectilinear figure.

14. *An Angle.* Two straight lines which intersect each other form an angle. The point of intersection is called the Vertex of the angle.

15. A *Right Angle* is formed when two straight lines meet and are perpendicular to each other.

16. An *Acute Angle* is less than a right angle.

17. An *Obtuse Angle* is greater than a right angle.

18. A *Triangle* is a plane surface bounded by three straight lines.

19. An *Equilateral Triangle* has all of its sides equal.

20. An *Isosceles Triangle* has two sides equal.

21. A *Scalene Triangle* has all of its three sides unequal.

22. A *Right-angled Triangle* has one of its angles a right angle.

23. An *Acute-angled Triangle* has three acute angles.

24. An *Obtuse-angled Triangle* has one obtuse angle.

25. The *Apex of a Triangle* is the upper extremity. Sometimes called the vertex.

26. The *Hypothenuse* is the longest side of a right-angled triangle. It is opposite the right angle.

27. The *Base* is the bottom side of a triangle.

28. The *Vertex* is the point in any figure opposite to and furthest from the base.

29. The *Altitude* of a triangle is the length of a perpendicular from the apex to the base.

30. A *Quadrangle* or *Quadrilateral* is a figure of four sides and has particular designations as follows:

Parallelogram, having its opposite sides parallel.

Square, having length and breadth equal.

Rectangle, all its angles are right angles.

Rhombus or *Lozenge* has equal sides but its angles are not right angles.

Rhomboid has only its opposite sides equal. Its angles are not right angles.

Trapezium, has unequal sides.

Trapezoid, only one pair of opposite sides are parallel.

Gnomon. The space included between the lines forming two parallelograms, of which the smaller is inscribed within the larger so as to have one angle in each common to both.

31. *Polygons* are plane figures having more than four sides and are either regular or irregular according as their sides and angles are equal or unequal. They are named from the number of their sides or angles, thus:

Pentagon, five sides	*Ocatgon,* eight sides	*Undecagon,* eleven sides
Hexagon, six sides	*Nonagon,* nine sides	*Dodecagon,* twelve sides.
Heptagon, seven sides	*Decagon,* ten sides	

32. A *Circle* is a plane figure bounded by a curved line every point of which is equidistant from the center.

33. A *Diameter* is a right line passing through the center of a circle or a sphere and terminated at each end by the periphery or surface.

34. An *Arc* is any part of the circumference of a circle.

35. A *Chord* is a right line joining the ends of an arc.

36. A *Segment* of a circle is any part of a circle bounded by an arc and its chord.

37. A *Radius* of a circle is a line drawn from its center to circumference.

38. A *Sector* is any part of a circle bounded by an arc and its two radii.

39. A *Semicircle* is half a circle.

40. A *Quadrant* is a quarter of a circle.

41. A *Zone* is a part of a circle included between two parallel chords.

42. A *Lune* is the space between the intersecting arcs of two eccentric circles.

43. A *Secant* is a line running from center of circle to extremity of tangent of arc.

44. *Cosecant* is secant of complement of an arc, or line running from center of circle to extremity of co-tangent of arc.

45. *Sine* of an arc is a line running from one extremity of an arc perpendicular to a diameter passing through the other extremity, and sine or angle is sign of arc that measures that angle.

46. *Versed Sine* of an arc or angle is part of diameter intercepted between sine and arc.

47. *Cosine* of an arc or angle is part of diameter intercepted between sine and center.

48. *Coversed Sine* of an arc or angle is part of secondary radius intercepted between cosine and circumference.

49. *Tangent* is a right line that touches a circle without cutting it.

50. *Cotangent* is tangent of complement of arc.

51. Circumference of every circle is supposed to be divided into 360 equal parts termed *Degrees;* each degree into 60 minutes and each minute in 60 seconds.

52. *Complement* of an angle is what remains after subtracting angle from 90°.

53. *Supplement* of an angle is what remains after subtracting angle from 180°.

54. *Vertex* in Conic Sections is the point through which the generating line of the conical surface always passes.

55. *Measure* of an angle is an arc of a circle contained between the two lines that form the angle and is estimated by number of degrees in arc.

56. *Segment* is a part cut off by a plane parallel to the base.

57. *Frustum* is part remaining after segment is cut off.

58. *Perimeter* of a figure is the sum of all its sides.

59. *Circular Cylinder.* A figure formed by revolution of a right-angled parallelogram around one of its sides.

60. *Prism.* A figure whose sides are parallelograms and ends equal and parallel.

61. *Wedge,* a prolate triangular prism.

62. A *Triangular Prism* has triangular bases.

63. A *Quadrangular Prism* has quadrilateral bases.

64. A *Pentagonal Prism* is one whose bases or ends are pentagons.

65. A *Hexagonal Prism* has hexagons for bases.

66. A *Cube* is a prism whose six faces are all squares.

67. An *Elliptical Cylinder* is one whose bases are ellipses.

68. An *Oblique Cylinder* is one whose curved surface is inclined to its bases.

69. A *Cone* is a round solid with a circle for its base, and tapering uniformly to a point at the top called the apex.

70. A *Right Cone* is one in which the perpendicular from the apex passes through the center of the base. This perpendicular is called the axis of the cone.

71. An *Oblique* or *Scalene Cone* has its axis inclined to the plane, of its base.

72. A *Truncated Cone* is one whose upper part is cut off by a plane parallel to its base.

73. A *Pyramid* is a solid with a straight-sided base and triangular sides terminating in the apex. Pyramids are named by the forms of their bases as triangular, quadrangular, pentagonal, · hexagonal, septagonal, etc.

74. A *Right Pyramid* has a regular polygon for a base and its axis is perpendicular to and passes through the center of the base.

75. A *Polyhedron* is a solid bounded by plane figures.

76. A *Tetrahedron* is bounded by four equilateral triangles.

77. *Hexahedron* is bounded by six squares, a cube.

78. *Octrahedron* is bounded by eight equilateral triangles.

79. *Dodecahedron* is bounded by 12 pentagons.

80. *Isocahedron* is bounded by 20 equilateral triangles.

81. A *Conic Section* is formed by the intersection of a cone and a plane. The different conic sections are the *triangle, circle, ellipse parabola* and *hyperbola.*

82. A *Triangular Section* is cut from a cone by a plane through the axis perpendicular to the base.

83. *Ellipse.* This section is formed when the cone is cut by a plane oblique to its opposite elements. An oblique section through a circular cylinder is also an ellipse.

84. *Parabola.* When a cone is cut by a plane parallel to one of its elements, the section is a parabola.

85. *Hyperbola.* When a cutting plane makes an angle with the base greater than the angle made by an element, the section is a hyperbola. A plane parallel to the axis passed on either side of it will give a hyperbola.

86. *Sphere.* A solid, the surface of which is at a uniform distance from the center.

87. *Ungulas.* Cylindrical ungulas are frusta of cylinders; conical ungulas are frusta of cones.

88. *Concave* means hollow, curved inwardly the opposite of *Convex*, which curves outwardly, bulging. A sphere has a convex curve.

CHAPTER III

FREEHAND LETTERING AND GEOMETRICAL DRAWING

ONE of the important parts of a commercial working drawing is the lettering including figures.

If a drawing is well made but poorly lettered and dimensioned, it looks bad. If the lettering is good the drawing will have a good appearance even if it has defects in its construction.

All lettering and figuring must be placed on a drawing freehand. Therefore, it is necessary to give all the time and effort required to become as proficient as possible in the making of good freehand letters and figures.

In Chapter 2, Article 22 specifies the style of letters to be used in this course. This style has been selected for its maximum of legibility considering its comparative ease of construction.

Heretofore it has been the custom of the writer when teaching mechanical drawing to give a course in freehand lettering before any drawing. In theory, this is good practice because it enables the student to letter his drawings as soon as they are made and to do so in a fairly creditable manner. It has been found, however, that to most beginners freehand lettering comes hard at first, and to interpolate a drawing occasionally between the plates of letters has proven to be a wise procedure.

The following method of teaching freehand lettering for use on machine or other working drawings is simple, brief and produces most excellent results. This method was introduced by the writer at Armour about ten years ago and it was probably the beginning of such a method in Chicago at least.

The value of freehand lettering cannot be emphasized too much. It is very desirable that a draftsman should be able to letter and figure his drawing with a plain, neat, properly formed, well-made freehand letter, of maximum legibility and comparatively easy to make.

Many " ads " for draftsmen these days require that applicants must be good freehand letterers, so it is well worth while to spend all the time and effort necessary to become a good letterer and figurer.

Straight-line Letters. The straight-line letters are taken up first because they are comparatively easy to make. All the letters of the alphabet both straight and curved will be drawn 6 spaces high, that is $\frac{3}{4}''$ (since the cross-section pad is divided 8×8).

This large letter is used at first to teach the *form* and *proportion* of the letter, together with the proper slope suitable for the height.

PLATE I. 4 HOURS

Before beginning to draw the letters, prepare the 4H pencil, as described in Article 4, Chapter 1, page 7, and illustrated by Figs. 9, 10 and 11.

This plate will consist of two sets of the straight-line letters of the alphabet. The second set must show a decided improvement over the first, see Fig. 7.

When ready to begin drawing the letters proceed as follows: Place the pad on the drawing board in front of you, the long way from left to right, and sitting in an easy position, holding the pencil not too tightly in the fingers, with the elbow close to the side, locate the lowest point of the letter " *I* " 12 spaces from the top of the pad and 7 spaces from the left-hand edge of the sheet. Then from that point 3 spaces to the right and 6 spaces up locate the highest point of this letter and mark it with the point of the pencil. Then place the point of the pencil just over this highest point and without touching the paper at first, draw down towards the lowest point and repeat the motion a few times without making any mark on the paper.

This motion is for the purpose of obtaining direction. Then gradually lower the pencil point and touch the paper lightly while drawing the letter. To make sure that the line of the " *I* " is straight, take up the pad and look at the line edgewise, you will easily detect any crookedness or curve. If it should not be quite straight, erase it gently with the Emerald pencil eraser, Fig. 30, Article 16, page 20, and try again.

But if you find it straight, then proceed to draw the letter
" **L** " in a similar manner.

Locate the lowest left-hand point of the " **L** " $3\frac{1}{2}$ spaces to
the right of the lowest point of the " **I**," then 3 spaces to the right
again for the slope, then up 6 spaces for the height and mark
the upper point of the " **L**."

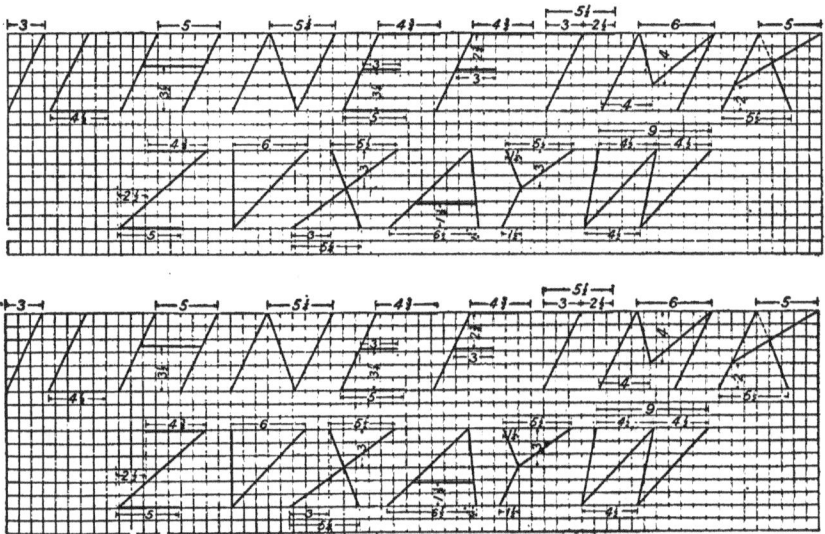

Fig. 38.

Next draw the down stroke of the " **L** " as was done in drawing the " **I**."

Then draw the bottom line from left to right $4\frac{1}{2}$ spaces,
and so on for all the straight-line letters according to their form
and proportions given in Fig. 38.

When all the letters in Plate 1 have been drawn and approved
the student should sign his name in the lower right-hand corner
together with plate number, the current date and also the time
taken to make the plate, counting the gross time, preparation,
drawing, etc.　(For example)

PLATE 1
3 HOURS, JULY 21, 1917.
JAMES GRISWOLD.

The plate will then be signed and recorded and returned to the student to be kept flat in manilla laboratory covers until all the lettering plates are finished when they are to be bound together in the cover plates with brass binders and a title lettered on the outside of the cover as follows:

Mechanical Drawing

FREEHAND LETTERING

PLATES 1, 3, AND 5 TO 10, INCLUSIVE

JULY 21, 1917

HAROLD DALE.

31. Geometrical Drawing. The problems in geometrical drawing are given for several reasons.

1. To teach the use of drawing instruments and at the same time learn the methods of construction of those problems in practical geometry that are the most useful in mechanical drawing, and to impress them upon the mind of the student so that he may readily apply them in practice.

2. To emphasize accuracy of construction. All dimensions should be laid off carefully, correctly and quickly in the manner directed in Article 11, page 17.

3. To impress upon the student by means of these drawings that straight lines joining arcs should be made exactly tangent so that the joints cannot be noticed.

It is the little things like these that make or mar a drawing, and if attended to or neglected they will make or mar the draftsman. The constant endeavor of the student should be to make every drawing he begins more accurate, quicker and better in every way than the preceding one.

A drawing should never be handed in as finished until the student is perfectly sure that he cannot improve it in any way whatever, for the act of handing in a drawing is the same, or should be the same, as saying " This is the best that I can do;" " I cannot improve it;" " it is a true measure of my ability to make this drawing."

32. Straight-line Problems in Geometrical Drawings. Only those straight-line problems that require the use of the

compass and spacer have been selected for drawing exercises in
this plate. Other straight-line problems that can be drawn
directly with triangle and T-square such as perpendiculars to
given lines, lines parallel to one another, etc.

PLATE 2. 6 HOURS

To prepare for drawing the problems in Plate 2.

1st. Use sheet of cream detail paper referred to in Article
20, page 22. Lay it on the board close to the top and make it
even with the T-square held rigidly against the left-hand end
of board with the left hand. When square with T-square and
board insert thumb tack in each upper corner. If the paper
curls up at first at the lower corners insert a thumb tack at each
of the lower corners also. After a while when the paper will lie
flat without the lower thumb tacks take them out. They should
be removed anyway when drawing with the T-square at the
bottom of the sheet.

2d. Get out the 4H and 6H pencils, pencil compass, dividers,
the two triangles, emerald eraser and pencil pointer.

3d. Put straight-line pencils and compass pencil in good order
as directed in Article 4, page 7.

4th. Draw border line on the drawing paper as follows:
Take the scale shown in Fig. 24, page 17, and using the edge of
$\frac{1}{16}$ths, lay off $\frac{1}{2}''$—that is half an inch—from the top and bottom
edges of the paper, and with the T-square and 4H pencil held as
illustrated in Figs. 3 and 4, draw straight lines from left to right,
leaning very lightly on the pencil, so as not to form a groove in
the paper. Then lay off $1\frac{1}{2}''$ from the left-hand edge at the top
and bottom and using the T-square as a straight edge, draw the
left-hand border line. Next lay off $\frac{1}{2}''$ from the right-hand edge
of the paper about $2''$ from the top and bottom and using the
T-square as before draw the right-hand border line.

5th. Lay out the sheet inside the border lines in 16 equal
parts, as follows. Divide the long space between the left- and
right-hand border lines which is equal to $18''$ by 4 making $4\frac{1}{2}''$.
Set the large dividers, Fig 20, page 13, to $4\frac{1}{2}''$ and step off 4
spaces on the upper border line from left to right, and mark the

divisions with the point of the pencil. Through these points draw vertical lines with the T-square and largest triangle held

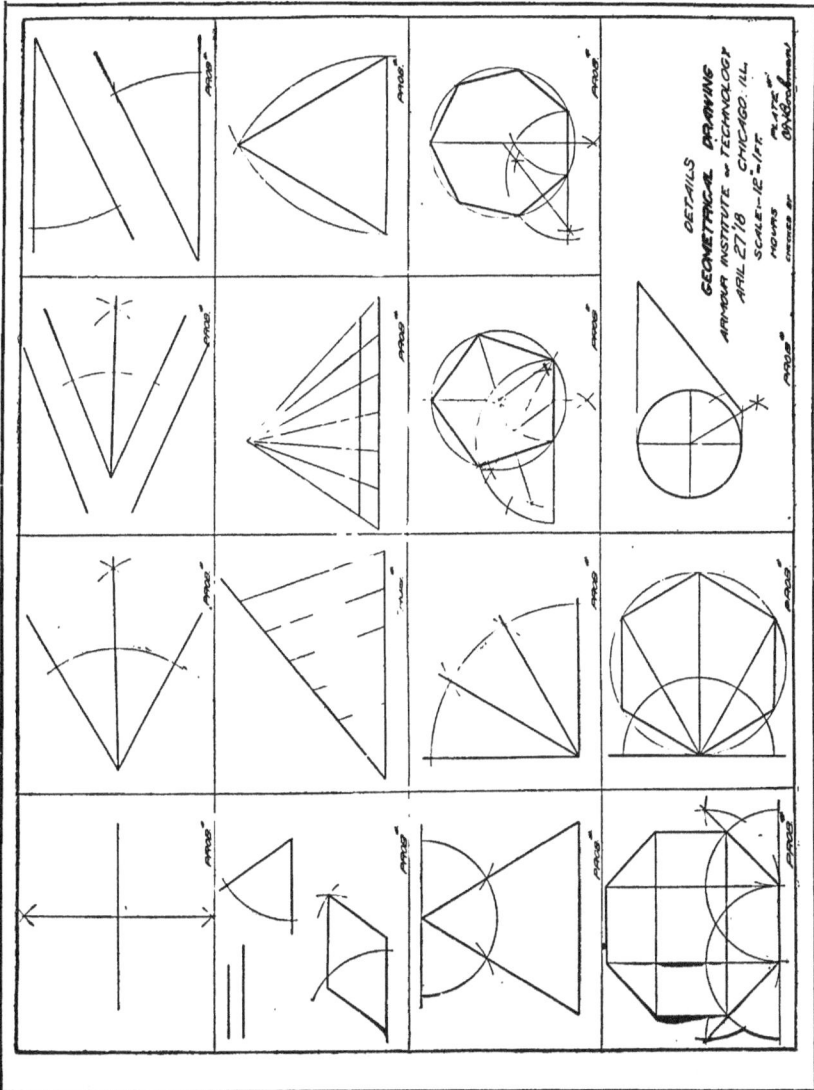

FIG. 39.

in a similar manner to that shown in Fig. 3. Divide the space between the top and bottom border lines into 4 equal spaces in the same way.

PROB. 1, FIG. 40.   *Bisect a Given Finite Straight Line.*

1st. Using the 6H pencil sharpened as described in Article 4, page 7, and the T-square, draw the given line AB $3\frac{1}{4}''$ long in the center of the first space on the sheet.   Letter the ends of the line A and B very lightly and carefully with the 4H pencil sharpened as directed in Section 4.

2d. Take the large compass and set it to a radius greater than half the length of AB, say $2''$ and placing the needle point at A and B, respectively, draw small arcs intersecting or cutting each other at E and F.

3d. With one of the triangles used as a straight edge, draw a

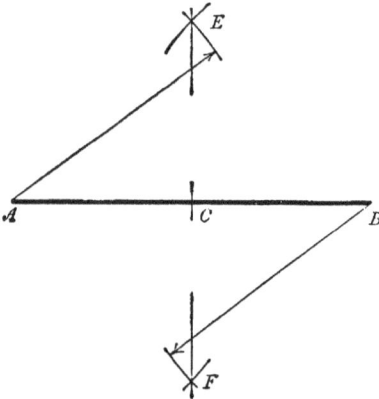

FIG. 40.

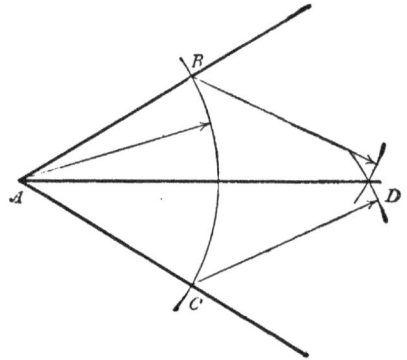

FIG. 41.

fine, narrow line through the points of intersection of the two small arcs at E and F, cutting AB at C.   At the point C the line AB is bisected, or divided into two equal parts.   An arc of a circle may be bisected in the same way.   Great care should be taken when drawing the line through E and F to make sure that it passes exactly through the point of intersection.   It is sometimes a good plan to locate the exact point of intersection with the point of the 4H pencil before drawing the line through EF. Leave this drawing in fine lines for the present and proceed next to make the drawing in Fig. 41.

PROB. 2, FIG. 41.   *Bisect a Given Angle by Dividing it into Two Equal Angles.*

1st. Assume the point A, Fig. 41, in the up-and-down center of second space on the sheet $\frac{1}{2}''$ from the left division line and mark it A. From A draw the given angle BAC making the legs AB and AC about $2\frac{1}{2}''$ long.

2d. With the point A as center and a radius about $1\frac{3}{4}''$ long draw an arc cutting the legs of the angle in the points C and B.

3d. With B and C as centers draw arcs cutting in D.

4th. Through the points A and D draw a line dividing the given angle into two equal angles.

PROB. 3, FIG. 42. *To Bisect an Angle when the Lines forming the Angle do not Extend to a Meeting Point.*

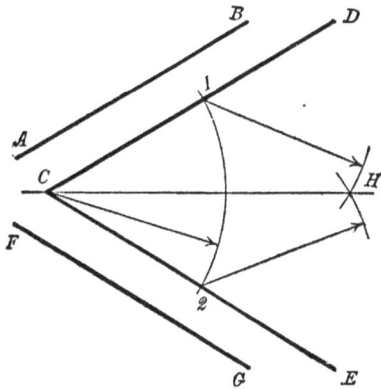 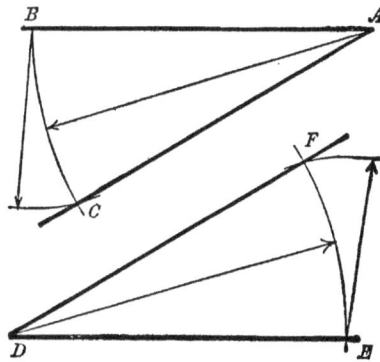

FIG. 42.   FIG. 43.

1st. Draw AB and FG, making any convenient angle with each other, locating them in the 3d space about as shown in Fig. 39.

2d. Draw the line CD parallel to AB and $\frac{1}{4}''$ from it. Draw the line CV parallel to FG, and produce them to meet at the point C.

3d. Bisect the angle DCV by Prob. 2. The bisector CH will also bisect the angle between the given lines AB and FG.

PROB. 4, FIG. 43. *From a point on a given line set off an angle equal to a given angle.*

Let BAC be the given angle and D the given point on given line DC.

1st. Draw the line AB 3″ long and 3″ above the bottom of space 4 and from the point A draw the line AC making an angle of 30° with AB using the 30°×60° triangle.

2d. With a radius of $3\frac{1}{4}$″ and the point A as center draw arc BC. From D as center and the same radius draw arc EF. With center E and radius BC cut arc EF in F.

3d. Through the D and F draw DF. The angle EDF is equal to the given angle BAC.

PROB. 5, FIG. 44. *Construct a rhomboid having adjacent sides*

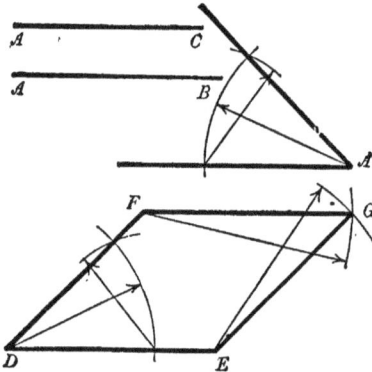
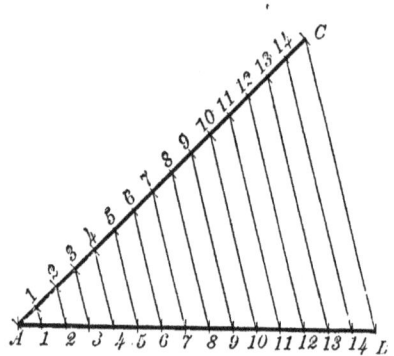

FIG. 44. FIG. 45.

equal to two given lines AB and CD, and an angle equal to a given angle A.

1st. Draw the given line AC $\frac{1}{2}$″ from the top of the 5th space and make $1\frac{3}{4}$″ long.

2d. Draw AB $\frac{1}{2}$″ below AC $2\frac{1}{2}$″ long.

3d. Draw the angle A equal to 30°.

4th. Draw the line DE $\frac{1}{2}$″ above bottom of space equal in length to line AB.

5th. With D on the line DE as center and any convenient radius, say 1″, describe an arc and make the angle at D equal to the given angle A. Make DF equal in length to AC.

6th. From F with a radius equal in length to the given line AB and from E with a radius equal to the given line AC describe arcs intersecting each other at G. Join FG and EG with straight lines completing the rhomboid as required.

It will do no harm at this point to remind the student that all the lines in these drawings must be made sharp and narrow—not faint and weak. This can be done by maintaining the pencil points in good order as already explained. Make an effort to obtain neatness and extreme accuracy of construction in all your work, especially in locating centers of arcs and in drawing lines through points of intersection.

PROB. 6, FIG. 45. *Divide a given line into any number of equal parts. A B is the given line. Divide it into 15 equal parts.*

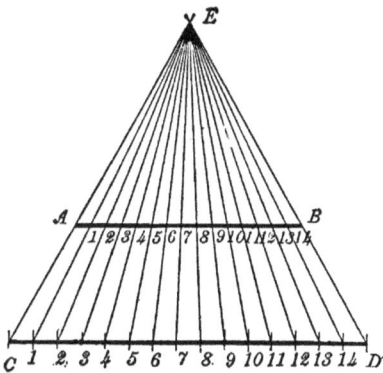

FIG. 46.

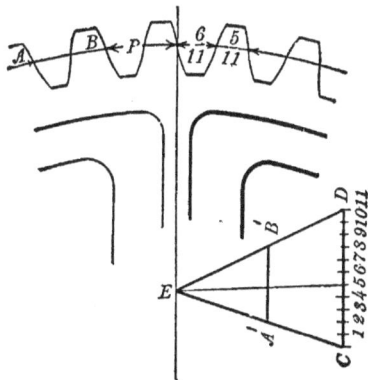

FIG. 45 a.

1st. Draw the given line AB $1\frac{1}{2}''$ above the bottom line of the space, and make $2''$ long.

2d. Draw another line CD parallel to AB and $1''$ below it. Make this $3\frac{1}{2}''$ long.

3d. From C on the line CD set off the number of equal parts into which the line AB is to be divided.

4th. Draw lines through CA and DB and produce them until they meet at E.

5th. Through each one of the points 1, 2, 3, 4, etc., draw lines to the point E, dividing the line AB into the required number of equal parts.

This problem is useful in dividing a line when the point required is difficult to find accurately—for example in Fig. 45a AB is the circular pitch of the spur gear, partly shown, which includes a space and a tooth and is measured on the pitch circle.

In cast gears the space is made larger than the thickness of the tooth, the proportion being about 6 to 5—that is, if we divide the pitch into eleven equal parts the space will measure $\frac{6}{11}$ and tooth $\frac{5}{11}$. The $\frac{1}{11}$ which the space is larger than the tooth is called the backlash.

Let $A'B'$, Fig 45a, be the pitch chord of the arc AB.

Draw CD parallel to $A'B'$ at any convenient distance and set off upon it eleven equal spaces of any convenient length.

Draw CA' and DB' intersecting at E.

From point 5 draw a line to E. This line divides $A'B'$ as required.

The one part $\frac{5}{11}$ and the other $\frac{6}{11}$.

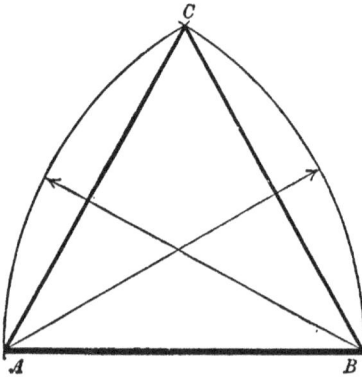

FIG. 47.

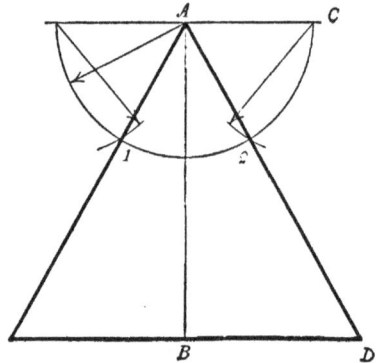

FIG. 48.

PROB. 7, FIG. 47. *Draw an equilateral triangle on a given base.*

1st. Draw the given base $\frac{1}{2}''$ above the bottom of the space and from the points A and B with AB as radius describe arcs cutting in C.

2d. Draw lines AC and BC. The triangle ABC is equilateral and equiangular.

PROB. 8, FIG. 48. *Construct an equilateral triangle of a given altitude AB.*

1st. Assume the altitude AB and at both ends of it draw lines perpendicular to it as CA, DB.

2d. From A with any radius describe a semicircle on CA, and with its radius cut off arcs 1, 2.

3d. Draw lines from A through 1 and 2 and produce them until they cut the base BD.

PROB. 9, FIG. 49. *Trisect a right angle.*

1st. Draw the line CB $2\frac{1}{2}''$ long, $\frac{1}{2}''$ above the bottom of the 9th space. With the right-angled triangle $30° \times 60°$ draw the perpendicular BA $2\frac{1}{2}''$ long, completing the given right angle.

2d. With B as center and BC as radius describe the arc AC.

3d. With the same radius and centers A and C cut the arc AC in points 1 and 2. Draw lines from 1 and 2 to B. These lines divide the right angle into 3 equal angles.

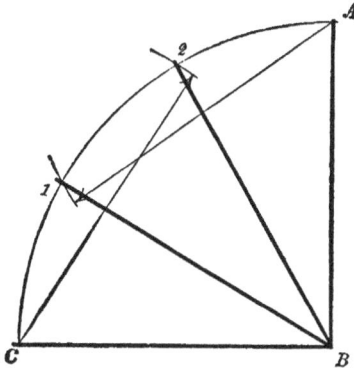

FIG. 49.

PROB. 10, FIG. 50. *Draw a pentagon on a given side AB.*

A pentagon is a five-sided figure and is often met with in practice.

1st. Draw AB $1\frac{3}{4}''$ long $\frac{1}{4}''$ above the bottom of the space and produce it to the left.

2d. With B as center and radius BA, draw the semicircle $C2A$.

3d. With center A and same radius draw arc BD, cutting first arc in the point D. Bisect AB in E and through it draw ED, which will be perpendicular to AB.

4th. Bisect arc BD in the point F and draw EF, then with center C and radius EF cut off arcs $C1$ and $1-2$ on the same semicircle.

5th. Draw line $B2$; it will be a second side of the pentagon.

Bisect it and draw a line through the point of bisection perpendicular to $B2$.

The perpendiculars through AB and $B2$ cut in G, which is the center of the circumscribing circle of the pentagon.

PROB. 11, FIG. 51. *Construct a heptagon on a given side AB.*
A heptagon is a seven-sided figure.

1st. Draw $AB\ \frac{1}{4}''$ above the bottom of the space $1\frac{1}{4}''$ long and produce it to the left.

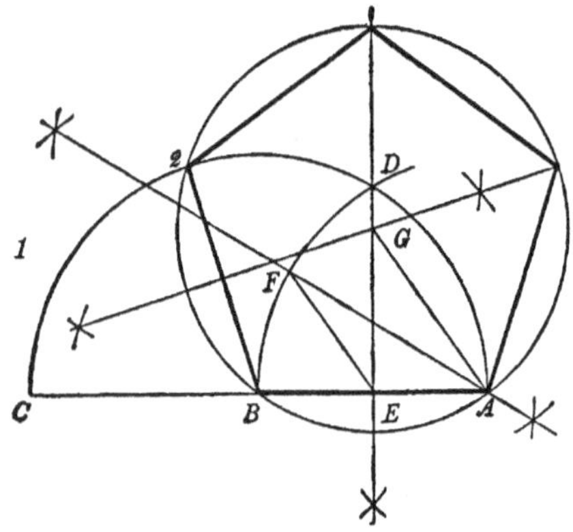

FIG. 50. 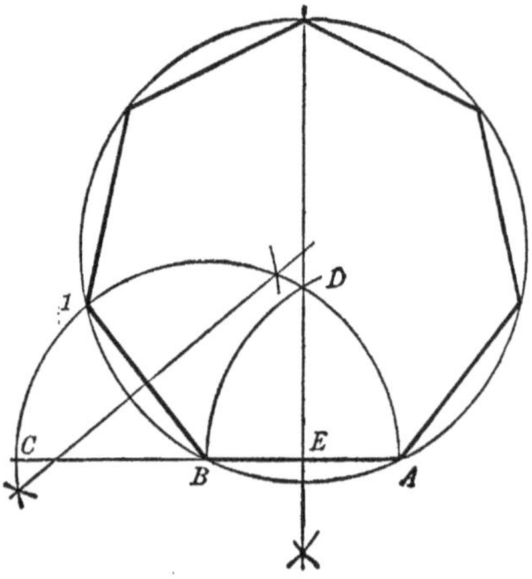   FIG. 51.

2d. With center B and radius BA describe a semicircle, and with A as center and same radius draw arc cutting the semicircle in D.

3d. Bisect AB in E and draw the perpendicular ED.

4th. With C as center and ED as radius, cut off arc $C1$ on the semicircle and draw $B1$; it is a second side of the heptagon.

5th. Bisect $B1$ and obtain the center circumscribing circle and finish drawing the heptagon as in the preceding problem.

These operations must be very carefully made to obtain an accurate construction of the figure.

PROB. 12, FIG. 52. *Construct a regular octagon on a given line AB.*

1st. Draw $AB\ 1\frac{1}{8}''$ long $\frac{1}{2}''$ above bottom of space and extend in both directions.

2d. Erect perpendiculars at A and B.

3d. With centers A and B and radius AB describe the semicircles CEB and AFD.

4th. Bisect the quadrants CE and DF in 1 and 2, then A1 and B2 will be two more sides of the octagon.

5th. At 1 and 2 erect perpendiculars 1–3 and 2–4 equal to AB. Draw 1–2 and 3–4.

6th. Make the perpendiculars at A and B equal to 1–2 or

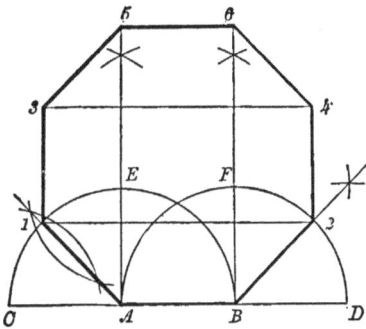

FIG. 52.

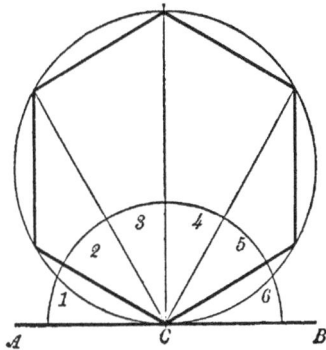

FIG. 53.

3–4, viz., A5 and A6. Complete the octagon by drawing 3–5, 5–6 and 6–4.

PROB. 13, FIG. 53. *Construct a regular polygon of any number of sides the circumscribing circle being given.*

1st. Draw a horizontal center line $1\frac{3}{4}''$ above bottom of space and at the center of it draw a vertical center line and at the point of intersection describe the given circle of $1\frac{1}{2}''$ radius.

2d. At any point of contact as C draw a tangent AB to the given circle.

·3d. From C with any convenient radius describe a semicircle cutting the given circle.

4th. Divide the semicircle into as many equal parts as the polygon is required to have sides, as 1, 2, 3, 4, 5, 6.

5th. Through each division draw from C lines to cut the circle in points of the hexagon. Join these points with straight lines and complete the figure.

PROB. 14, FIG. 54. *Draw a right line equal to half the circumference of a given circle.*

1st. Draw a vertical diameter AB 3″ long and $1\frac{3}{4}$″ from the left.

2d. Draw the horizontal center line in center of the space and describe the 3″ given circle.

3d. Draw AC perpendicular to AB and eqal to 3 times the radius of the circle.

4th. At B draw BE perpendicular to AB.

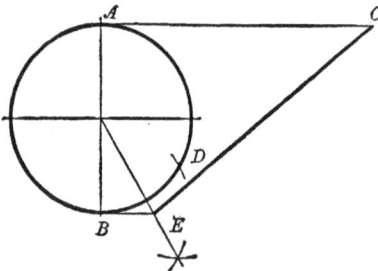 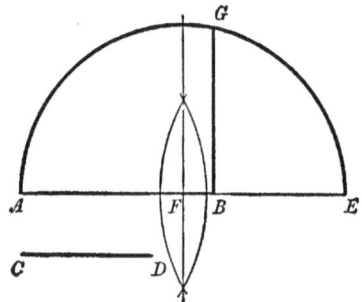

FIG. 54.   FIG. 55.

5th. With B as center and radius of the circle cut off arc BD, bisect it and draw a line from center of circle through the bisection, cutting the line BE in E.

6th. Join EC. Line EC will be equal to half the circumference of the given circle.

PROB. 15, FIG. 55. *Find a mean proportional to two given right lines AB and CD.*

1st. Draw AB 2″ long $1\frac{1}{4}$″ above the bottom line of the space and locate the point A $\frac{1}{2}$″ from the left.

2d. Draw CD $1\frac{1}{2}$″ long, $\frac{1}{2}$″ below AB.

3d. Extend the line AB to E, making BE equal to CD.

4th. Bisect AE in F and from F with radius FA describe a semicircle.

5th. At B where the two given lines join erect a perpendicular to AE cutting the semicircle in G.

BG will be a mean proportional to CD and AB.

Important

When the problems in Plate 2 have been completed in fine pencil lines and approved, the plate should be cleaned with the art gum and then all the given and required lines strengthened with the 4H pencil sharpened as directed in Art. 4.

The problem numbers should then be neatly lettered in the lower right-hand corner of each space. The height of the letters to be $\frac{3}{32}''$ and the numbers $\frac{5}{32}''$.

A title like the standard title, Fig. 35, page 24, should then be drawn carefully in pencil and when all the lettering has been approved, the border line and all lettering should then be inked in, and the drawing signed and recorded.

Since, as has been already emphasized the lettering and figures must be neat and well made, it will be best to defer lettering the early drawing plates until after the 4th plate of lettering has been finished. The student by that time should be able to make fairly good lettering and the lettering on his drawing plates will be done with greater facility. Every succeeding plate will be another opportunity to practice his lettering and to show how much he has profited by his study and practice of the first four letter plates.

Making the Title

1st. Draw all guide lines fine and narrow.

2d. Letter on a separate paper the longest line in the title as " Armour Institute of Technology," and measure its length then from $\frac{1}{16}''$ inside the right-hand border line, lay off that length to the left and letter that line first, when done, bisect it and draw a light line through the point of bisection that will be the center line of the title and the remainder of the title should be balanced with reference to that center line.

3d. Submit the title for approval.

4th. Ink the title and make the letters in the words of the main title as " Geometrical Drawing," very heavy so that they will be emphasized and stand out from the rest. Do not forget that numbers combined with letters are to be taller than the letters.

The next plate of lettering will consist of the curved letters of the alphabet and since they are more difficult to make than the straight-line letters more pains will be required to obtain the desired result.

PLATE 3. FIG. 56. 6 HOURS

These letters are to be drawn 6 spaces high and it will be seen from Fig. 56 that two sets of the curved letters of the alphabet are required. The second set ought to show a decided improvement over the first. The curves should be carefully analyzed as to their true form and proportion given in Fig. 56.

Submit each letter as soon as drawn for criticism and correction. In this way the art of obtaining good truly formed letters will be acquired much quicker than if several letters are made before having them examined and criticized. The curves of all the letters in this plate are similar so if the first letters are well made and the curves understood the succeeding letters will be easier to make and show improvement.

With a proper appreciation of the extreme importance of being able to make good freehand lettering and a close application to work this plate should be finished within the number of hours allowed for it.

Prepare to draw these curved letters in a similar manner to the way described in Article 32.

Using the cross-section pad and 4H pencil as before begin drawing the letter " U " as follows:

First draw the two guide lines marked A and B, giving them the required slope of three spaces.

Beginning at the point 1, draw the narrow curve downward until tangent to the bottom line at 3. Lift the pencil and commence the downward stroke of the larger curve at

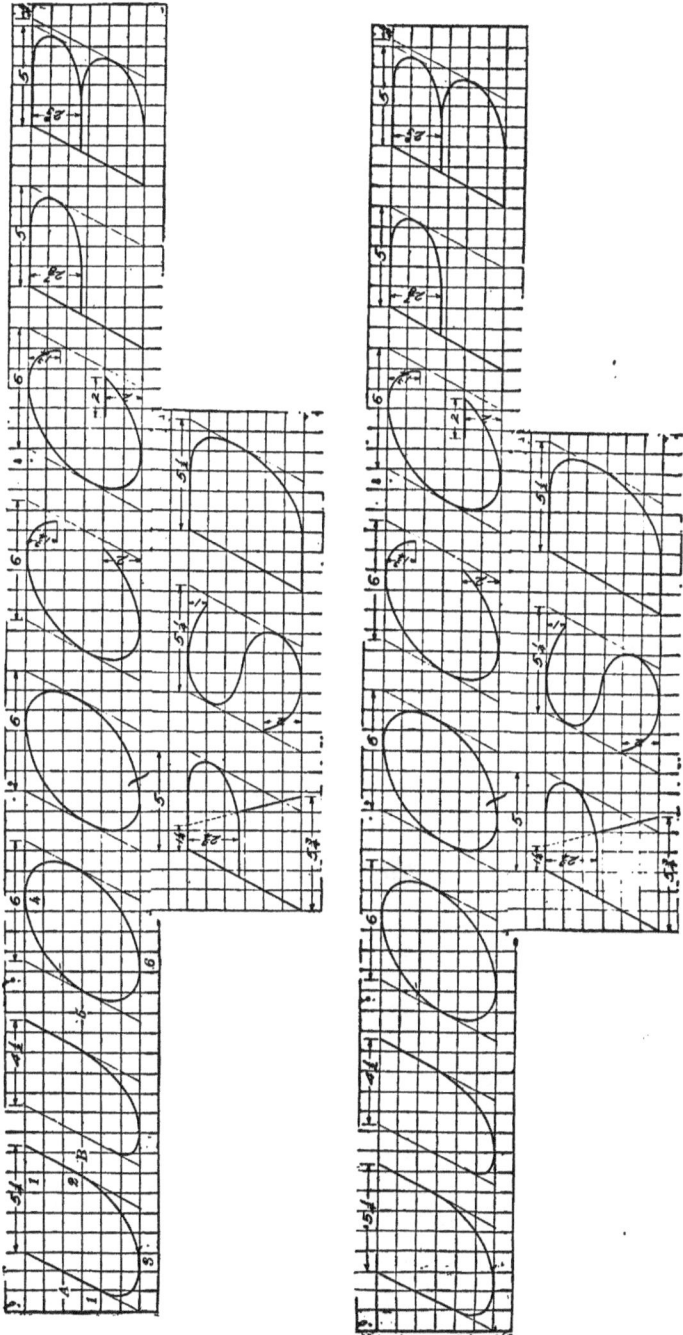

FIG. 56.

the point 2, drawing a smooth, even curve until tangent at the point 3.

Press very lightly with the pencil and try to obtain a narrow, smooth line.

The " J " is quite similar to the " U," it is a little narrower, therefore, the curves are slightly shorter.

When ready to draw the curve of the " O " begin at point 4 and draw down and to the left, making it tangent to the guide lines at the points 5 and 6.

Then from 4 again, draw to the right and down to 6.

By studying carefully where the curves cross the vertical and horizontal lines of the cross-section paper it should not take long to obtain the true form.

Note that the curve in the upper right-hand corner of the " P," " B," " R " and " D " is the same kind of a narrow curve in all of them. The lower curve in these letters is of the same nature as the lower right-hand curve in the others.

· It should be noticed in the letter " S " that the upper half is a little smaller than the lower half and that the top and bottom curves are flatter than the curves of any of the other letters.

37. Geometrical Drawing. *Continued.*

PLATE 4, FIG. 57.   8 HOURS

This plate is to consist of 12 geometric problems involving the use of the compasses as well as the triangles and T-square. Included are figures of the conic sections and other curves.

PROB. 16, FIG. 58. *Find the center of a given arc.*

1st. Draw an arc with a $2\frac{1}{2}''$ radius. Center of arc to be $\frac{1}{4}''$ above bottom line of space.

2d. Draw chords AB and BC and bisect them.

3d. Produce bisectors to meet in center required.

PROB. 17. FIG. 59. *Draw the involute of a circle.*

1st. Draw the given circle with $\frac{1}{2}''$ radius center $2\frac{1}{4}''$ above bottom line and $3\frac{1}{4}$ from left line of the space.

2d. Divide the circle into 12 equal parts with the 2 triangles and T-square and draw radii.

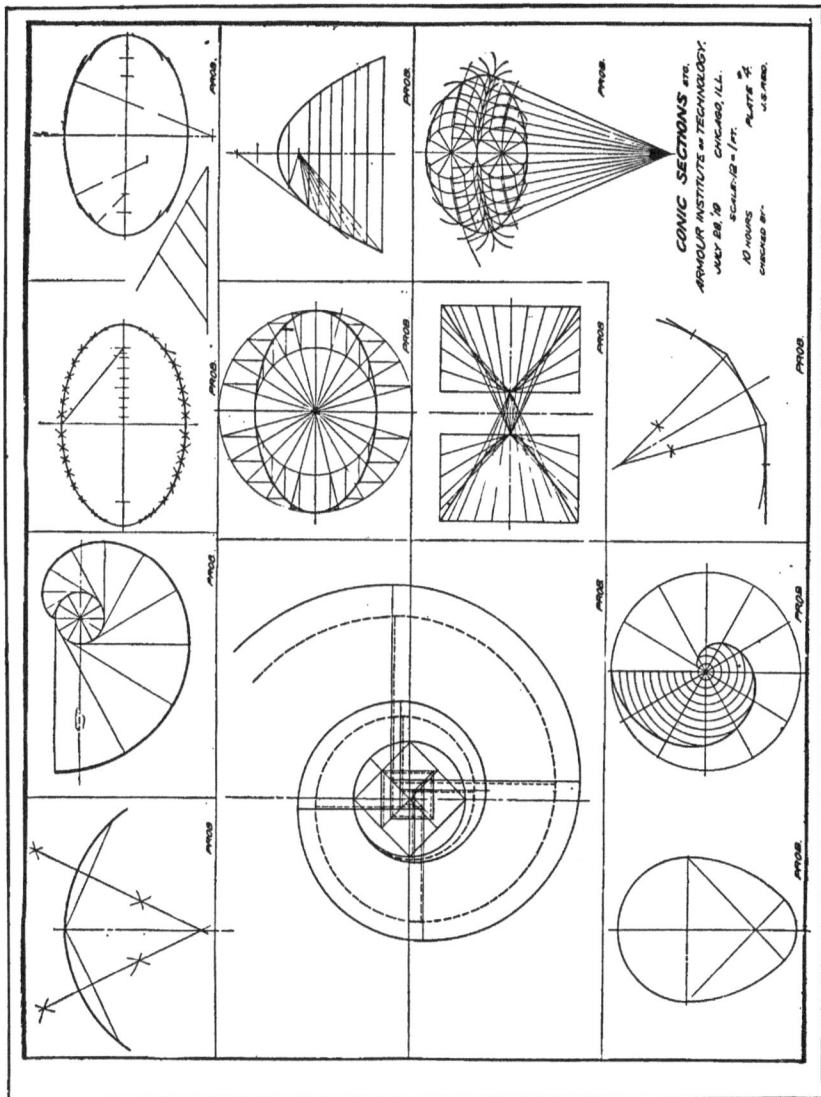

Fig. 57.

3d. Draw tangents at right angles to the radii.

4th. On the tangent to radius 1 lay off a distance equal to one of the parts into which the circle is divided.

5th. On each of the tangents set off the number of parts corresponding to the number of the radii. Tangent 12 will be the circumference of the circle unrolled, and the curve

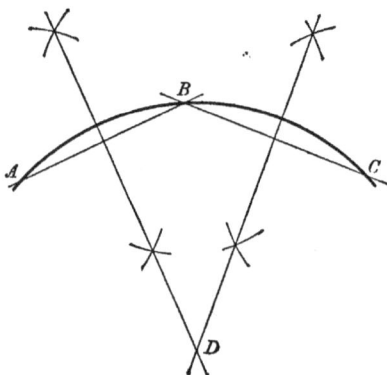

FIG. 58.

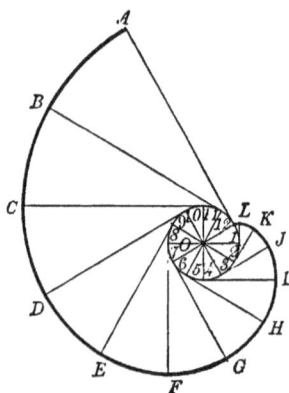

FIG. 59.

drawn through the extremities of the other tangents will be the involute required.

The curve may be drawn with arcs of circles, thus: Take 1 as center and 1–12 as radius and draw arc 12–L, then with 2 as center and 2–L as radius draw arc L–K and so on around the circle.

PROB. 18. FIG. 60. *Draw an ellipse with a given major axis AB and minor axis CD.*

The ellipse is a conic section. See Fig.

1st. Draw the major axis $3\frac{1}{2}''$ long in the vertical center of the next space and the minor axis $2\frac{1}{4}''$ long $2''$ from the left division line.

2d. With center C and radius AE cut AB in F and F' the foci.

3d. Divide EF into any convenient number of parts, say 7 as 1, 2, 3, 4, etc.

4th. With F as center and $A1$ as radius, draw a short arc above and below the horizontal center line about where the curve of the ellipse would be likely to come and F' as center, same radius draw small arcs on the other side of the minor axis. Then with $B1$ as radius draw from F and F' as centers arcs cutting the former arcs in points in the curve of the ellipse. Repeat the process from 1 to 7. To obtain a point in the curve between R and B take an extra division between 6 and 7 for radii.

5th. Draw the curve through the points of intersection of the arcs as at S and R, etc., either with the irregular curve

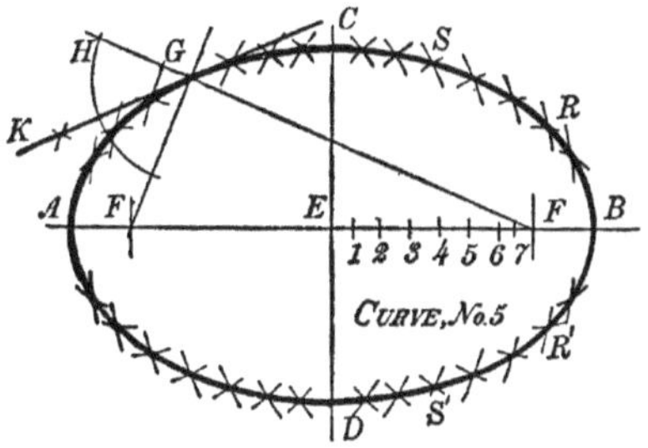

FIG. 60.

shown in Fig. 28, or by arcs of circles found in the manner described in the next problem.

The above method of finding points in the curve of the ellipse is theoretically correct, but the following method gives an approximation almost as close as can be drawn for small ellipses.

PROB. 19. FIG. 61. *Given the major and minor axes draw an ellipse by the following close approximent method.*

1st. Draw major and minor axes as before locating them about as shown in Plate 4.

2d. Draw any convenient angle like that shown in No. 2 and with radius equal to half the minor axis and point H as center draw arc LM and with radius equal to half the major axis and same center draw arc NO.

3d. Draw the right line LO and through M and N, respectively, draw lines MK and NP parallel to LO.

4th. With C and D as centers and radius $= HP$ mark the points 1 and 1' on the minor axis and from A and B with a distance equal to HK lay off the points 2 and 2' on the major axis. Then with centers 1, 1', 2, and 2' and radii 1D and 2A, draw arcs of circles as shown in Plate 4.

5th. To complete the ellipse between these arcs of circles use a piece of tracing cloth like that shown at T in Fig. 61, draw a narrow line in the center of it on the dull side of the cloth and puncture a small hole at G. From G lay off GF equal to the semi-minor axis and GE equal to the semi-major axis.

6th. Place the point F on the major axis and point E on the minor axes and move the strip of tracing cloth so that the point E is always in contact with the minor axis and point F with the major axis when the necessary points may be marked through the puncture at G with the sharp-pointed 4H pencil and the curve of the ellipse completed in this manner.

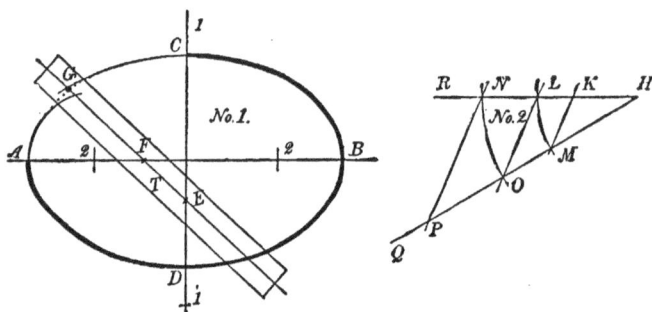

FIG. 61.

If tracing cloth is not at hand use a card or piece of stiff paper and mark the points G, F and E on its edge.

Instead of using the above method to complete the curve of the ellipse, an arc of a circle may be used which is a very close approximation to the correct curve. Add the half of the major axis to the half of the minor axis and divide by 2, that is $\dfrac{1\frac{3}{4}''+1\frac{1}{8}''}{2} = 2\frac{7}{8} \div 2 = 1\frac{7}{16}''$ the radius of the arc whose center may easily be found by trial.

PROB. 20. FIG. 62. *Given the same axes draw an ellipse by the following method.*

1st. With A as center and radii AB and AC describe circles. Draw any convenient radii as $A3$, $A4$, etc.

2d. Make 3-1, 3-4, etc. perpendicular to AB, and $D2$,

$E5$, etc., parallel to AB. Then 1, 2, 5, etc., are points on the curve. Draw the whole ellipse.

PROB. 21. FIG. 63. *Given the directrix BD and the focus C, draw a parabola and a tangent to it at the point 3.*

The parabola is a curve such that every point in the curve is equally distant from the directrix BD and the focus C. The vertix E is equally distant from the directrix BD and the focus C, that is CE, is equal to EB. Any

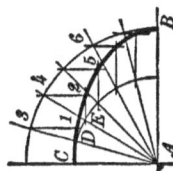

FIG. 62

line parallel to the axis is a diameter. A straight line drawn across the figure at right angles to the axis is a double ordinate, and either half of it is an ordinate. The distance from C to any point on the curve, as 2 is always equal to the horizontal distance from that point to the directrix. Thus $C1$ is equal to 1 $1'$, $C2$ to 2 $2'$, etc. To make the drawing:

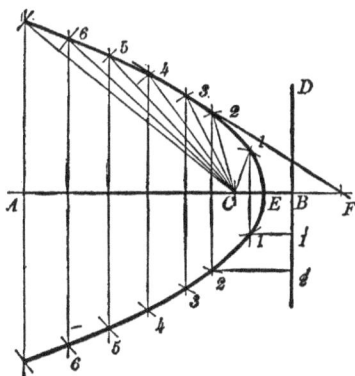

FIG. 63.

1st. Draw the axis AB in the center of the space, $3''$ long, and extend to F.

2d. At B draw BD the directrix at right angles to AB.

3d. Draw parallels to BD through any points in AB such as 1 1, 2 2, etc.

4th. Locate the focus C, $\frac{3}{4}''$ from B and bisect BC in E, the vertix.

5th. With C as a center and a distance equal to 1 $1'$, cut the parallel in the points 1 and 1. Then with the same center and distance equal to 2 $2'$ in the points 2 and 2 and so on for the other points in the curve.

6th. To draw the tangent at the point 3. Make EF equal to the distance of ordinate 3 3 from E. Draw the tangent through 3 F.

PROB. 22. FIG. 64. *Describe an ionic volute.*

When the volute is to be drawn to a given height. Divide the given height into seven equal parts, and through the point 3 draw 3, 3 perpendicular to AB. The eye of the volute the small circle NP should be made equal in diameter to one of the divisions on AB.

To describe the method of drawing the curves of the volute a small part to an enlarged scale will be drawn.

1st. In the center of 4 spaces draw the eye of the volute $2''$ diameter (see Plate 4) and draw the axes NN and PM.

2d. Inscribe the square $NPNM$. Bisect its sides and draw the square 11, 12, 13, 14. Draw its diagonals 11, 13 and 12, 14.

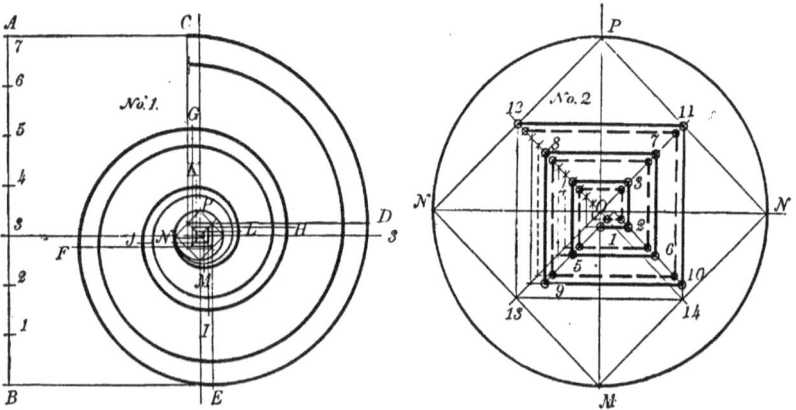

FIG. 64.

3. Divide the half diagonal o, 12 into 3 equal parts and subdivide each of these into 4 parts making 12 divisions in all from o to 12.

4th. At the second division from o drop a perpendicular to the diagonal 11, 13, and from where it cuts the diagonal draw a horizontal to cut the axis PM in the point 1 which is the first center of the outer curve of the volute.

5th. Through the point 1 draw a horizontal line. Through the 4th division on line o, 13 draw a horizontal line to intersect diagonal 11, 13 at point 3 from 3 drop a perpendicular to meet the horizontal through 1, at the point 2. Through 2 draw a line downward parallel to diagonal 12, 14. Point 2 is the second center of the volute outer curve. The other centers 3, 4, etc., are evident.

6th. From the first division on line o, 12 drop a perpendicular to meet diagonal 11, 13. Through the point of inter-

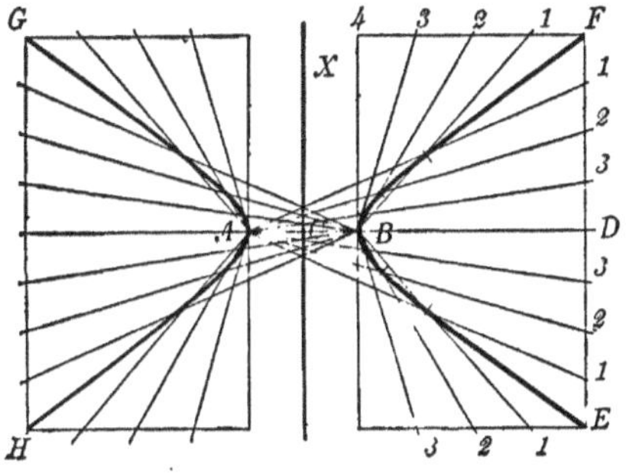

FIG. 65.

section draw a horizontal to cut the line through 1 parallel to diagonal 11, 13, in a point which is the first center of the inner curve of the volute shown as a broken line in Plate 4. The other centers will be found by following around the dotted squares.

PROB. 23. FIG. 65. *Draw an hyperbola having a given diameter AB, abscissa BD and double ordinate EF.*

1st. Draw the center line ABD about 4″ long and make AB equal to $\frac{3}{4}$″ in the center of the space.

2d. Draw FE $1\frac{1}{2}$″ from B and complete the rectangle $4FE$, making $F4$ parallel and equal to BD, and DF equal to $1\frac{1}{4}$″.

3d. Divide DF and $F4$ into the same number of equal parts, and from B draw lines to the points in $4F$, and from A draw lines to the points in DF.

4th. Draw the curve through the points where the lines correspondingly numbered intersect each other. Repeat below the other half of the curve as indicated.

PROB. 24. FIG. 67. *To draw the epicycloid and the hypocycloid.*

1st. Draw the arc BC with a radius equal to $3\frac{1}{2}$″, using two spaces on the drawing paper.

2d. In the most convenient position draw the line AF and

with a radius equal to $\frac{7}{16}''$ describe the generating circles FC and CG tangent to the directing circle BC at the point C.

3d. Divide the generating circles into any number or equal parts as 1, 2, 3, etc., and set off the length of these divisions from C on CB as e', d', c', etc.

4th. From A the center of the directing circle draw lines

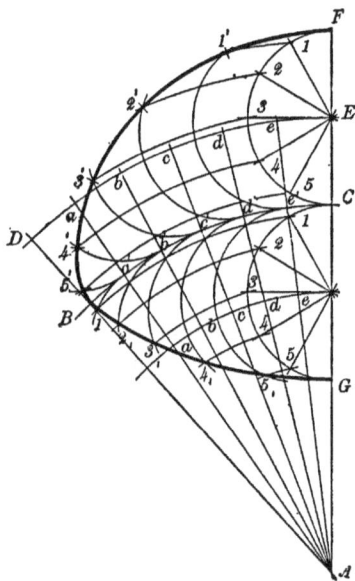

FIG. 67.

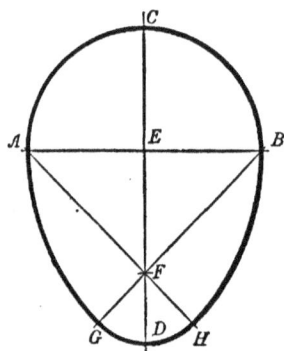

FIG. 68.

through e', d', c', etc., cutting the circles of centers in e, d, c, etc.

5th. From each of these latter points as centers describe arcs tangent to the directing circle CB.

6th. From center A draw arcs through the points of division on the generating circle, cutting the arcs of the generating circles in their several positions at the points 1', 2', 3', etc. 1, 2, 3, etc. These will be points in the two curves.

PROB. 25, FIG. 68. *To construct an oval the width AB being given.*

1st. Draw AB and CD at right angles to each other in the center of a space making AB equal to $2\frac{1}{2}''$ long.

2d. With E as center and EA as radius, draw the semi-circle ACB and cutting CD in F.

3d. From A and B draw lines through F, and from A and B as centers and AB as radius draw arcs cutting these lines in G and H.

4th. With F as center and radius FG describe the arc GH to meet the arcs AG and BH, and complete the curve of the oval.

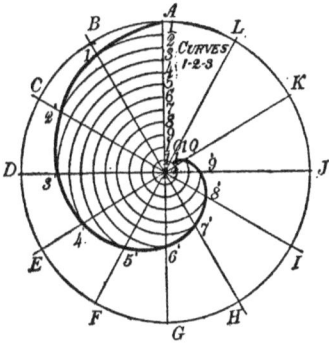

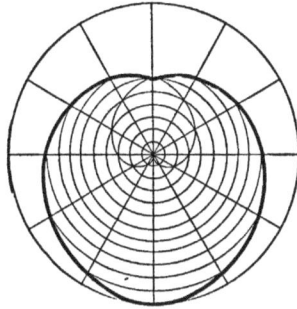

FIG. 69.   FIG. 70.

PROB. 26, FIG. 69. *Construct the Archimedes spiral of one revolution.*

1st. Describe a circle using the widest limit of the spiral as a radius, say $1\frac{1}{2}''$, and divide the circle into any number of equal parts, say 12.

2d. Divide the radius into the same number of equal parts as 1 to 12.

3d. From the center with radius 12 1, describe an arc cutting the radial line B in the point 1'.

4th. From the center continue to draw arcs from points 2, 3, 4, etc., cutting the corresponding radii C, D, E, etc., in points 2', 3', 4', etc., and through the points A, 1', 2', 3', etc., with the irregular curve draw the spiral.

Fig. 70. This figure shows the application of the Archimedes Spiral in the heart-shaped cam, which is used to change rotary motion into uniform reciprocating motion.

PROB. 27, FIG. 71. *Draw an arc of a circle tangent to two straight lines BC and CD given the mid position G.*

1st. Draw BC $2\frac{1}{2}''$ long $\frac{3}{4}''$ above the border line and CD $2\frac{1}{2}''$ long at an angle of 120° to BC.

2d. Draw CA the bisector of angle BCD, and EF, at right angles to CA through the point G.

3d. Bisect either of the angles BEF or EFD this bisector will intersect the center line CA at A the center of the required arc.

4th. From A draw perpendiculars $A1$ and $A2$, and with either as a radius and A as center, describe an arc which will be tangent to the lines BC and CD at the points 1 and 2.

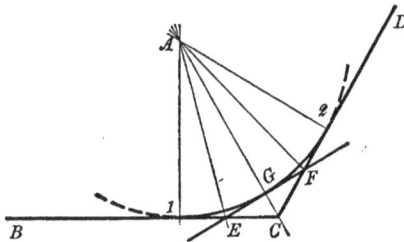

FIG. 71.

38. Freehand Lettering. Plate 5 will consist of the same style of letters drawn in Plates 1 and 3. In this plate they are to be drawn 3 spaces, 2 spaces and 1 space high.

1st. About $1\frac{1}{2}''$ from the top and left-hand edge of the cross-section pad begin the guide lines for the 3 space high letters. These guide lines should be fine, narrow, light lines drawn with 4H pencil sharpened as explained in Art. 7. The 3 space letters—that is 3 spaces high—should not be more than 1 space apart and the slope of the letters should be in the same proportion to the height as for the 6 space high letters, namely $1\frac{1}{2}$ spaces, in short, all dimensions of the 3 space high letters are to be made just half of those of the 6 space high letters.

2d. Draw the 2 space high letters using guide lines as before and make all dimensions and slope exactly one-third of the 6 space high letters. For the vertical spacing of the lines of lettering in this plate, see Fig. 72.

PLATE 5.   6 HOURS

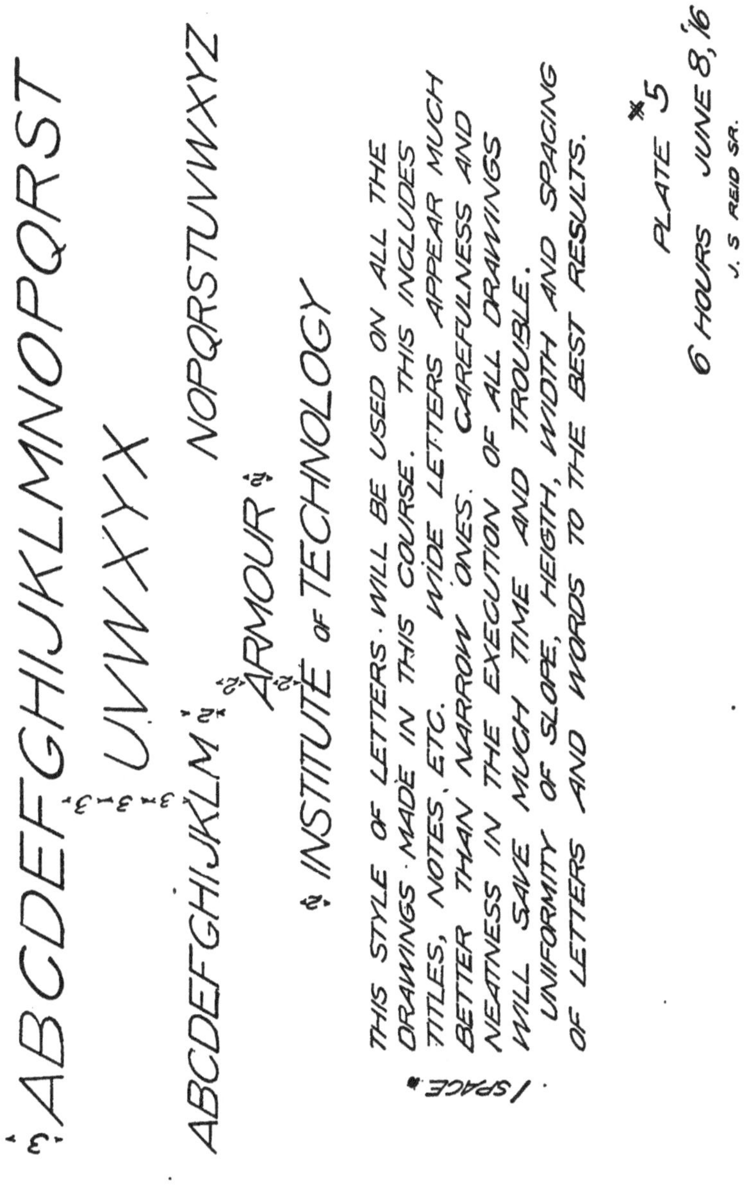

ABCDEFGHIJKLMNOPQRST
UVWXYX
NOPQRSTUVWXYZ

ABCDEFGHIJKLM

ARMOUR
INSTITUTE of TECHNOLOGY

THIS STYLE OF LETTERS WILL BE USED ON ALL THE
DRAWINGS MADE IN THIS COURSE. THIS INCLUDES
TITLES, NOTES, ETC.   WIDE LETTERS APPEAR MUCH
BETTER THAN NARROW ONES.   CAREFULNESS AND
NEATNESS IN THE EXECUTION OF ALL DRAWINGS
WILL SAVE MUCH TIME AND TROUBLE.
UNIFORMITY OF SLOPE, HEIGTH, WIDTH AND SPACING
OF LETTERS AND WORDS TO THE BEST RESULTS.

1 SPACE

PLATE 5
6 HOURS   JUNE 8,'16
J. S. REID SR.

FIG. 72.

3d. In drawing the 1 space high letters there are four rules to be observed. 1. Increase the **slope** to not less than three-quarters of the height. This is necessary because the small letters do not seem to slope as much as the large letters when the slope is made just half the height. 2. The **width** of the small letters must also be increased because wide letters 1 space high have a better appearance than narrow ones. 3. Open letters, that is, letters like E and F, should be placed close together and closed letters like H and I should be spaced farther apart. 4. Words should be placed 2 spaces apart for this size of letter. In Fig. 72 the lettering is by no means perfect. The student should examine each letter for form, proportion and spacing and improve upon them in all possible ways.

4th. When all the letters in this plate have been properly pencilled in, they should be inked with the ball-pointed pen. The inked lines of the letters should be of medium width and drawn smoothly with very little pressure on the pen.

PLATE 6. This plate will consist of the numerals as shown in Fig. 73, viz., one set of numbers 6 spaces high, one set 3 spaces high, one set 2 spaces high, together with fractions and 1 space high numbers.

1st. Locate the bottom of the set of 6 space high letters $1\frac{3}{4}''$ from the top of the sheet and $\frac{3}{4}''$ from the left-hand end. Draw the guide lines as indicated and pencil in the numbers with the 4H pencil sharpened to a conical point.

2d. Draw the set of 3 space high numbers.

Use guide lines and make the widths in the same proportion to the height as in the 6 space high numbers.

3d. Draw the large fractions, making the numerators and denominators 2 spaces high and the space for the dividing line 1 square high. Use guide lines and proportions as above. Care should be taken to obtain the proper slope for the whole fraction, as shown in the figure.

4th. Draw the 2 space high numbers and smaller fractions. Guide lines should be used for all numbers 2 spaces high and over.

PLATE 6

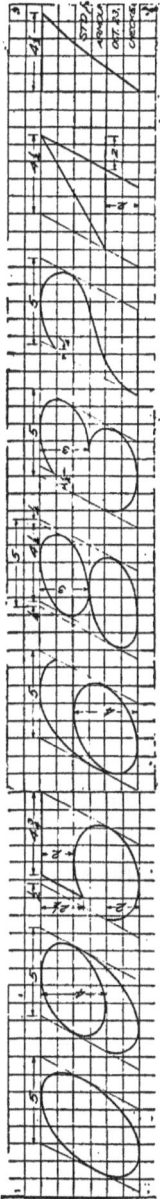

FIG. 73.

The 1 space high numbers should be drawn without guide lines. The 2 and 1 space numbers are to be inked after pencilling. The larger numbers are to be finished in pencil.

PLATE 7. This plate is to consist of 1 space high letters, all capitals, Gothic style, as shown in Fig. 74.

The directions given for drawing the small letters in Plate 5 will also apply here. Draw carefully in pencil first and then ink in with the ball-pointed pen.

PLATE 8. This plate is to consist of *lower case* letters 6 spaces high and arranged on the plate as shown in Fig. 75.

This style of letter is used by some draftsmen for notes on structural drawings, but the Gothic style, all capitals of the same height is the most popular. Finish in pencil. Ink title only.

PLATE 9 will consist of lower case letters, 1 space high.

The form and proportion will conform to the form and proportions given in Plate 8.

The slope of the letters in this plate should be about three-quarters of the height. Letter plate with 4H pencil and when approved, ink with Gillott pen No. 303.

PLATE 10 is to consist of Gothic capitals, 1 space high. Slope not less than three-quarters of the height.

Notice the extra space between paragraphs.

The numbers should be $1\frac{1}{2}$ times higher than the height of the letters. On all occasions when letters and numbers are combined the numbers should always be a half higher than the letters.

This being the last plate of lettering it should show a maximum of improvement over preceding plates. Pencil in with 4H and ink with ball-point No. 516. The lettering should be approved by the instructor before inking.

PLATE 7.   2 HOURS

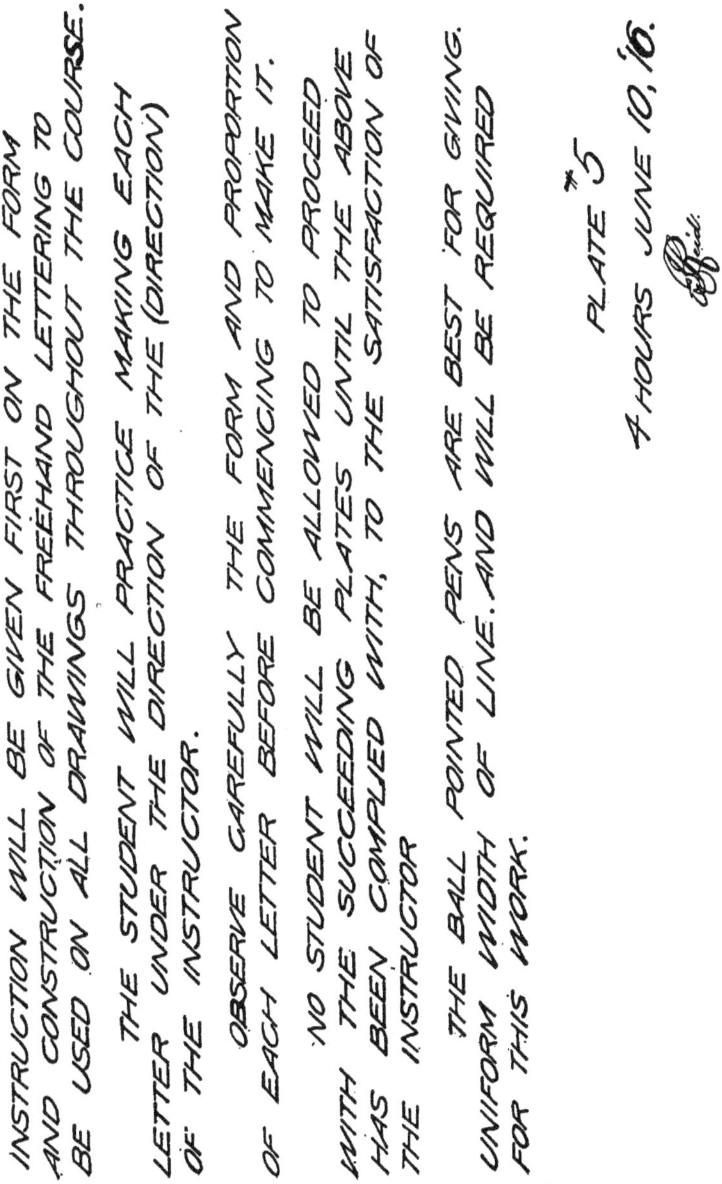

INSTRUCTION WILL BE GIVEN FIRST ON THE FORM AND CONSTRUCTION OF THE FREEHAND LETTERING TO BE USED ON ALL DRAWINGS THROUGHOUT THE COURSE.

THE STUDENT WILL PRACTICE MAKING EACH LETTER UNDER THE DIRECTION OF THE (DIRECTION) OF THE INSTRUCTOR.

OBSERVE CAREFULLY THE FORM AND PROPORTION OF EACH LETTER BEFORE COMMENCING TO MAKE IT.

NO STUDENT WILL BE ALLOWED TO PROCEED WITH THE SUCCEEDING PLATES UNTIL THE ABOVE HAS BEEN COMPLIED WITH, TO THE SATISFACTION OF THE INSTRUCTOR

THE BALL POINTED PENS ARE BEST FOR GIVING. UNIFORM WIDTH OF LINE. AND WILL BE REQUIRED FOR THIS WORK.

PLATE *5

4 HOURS JUNE 10, '16.

FIG. 74.

Plate 8.   2 Hours.

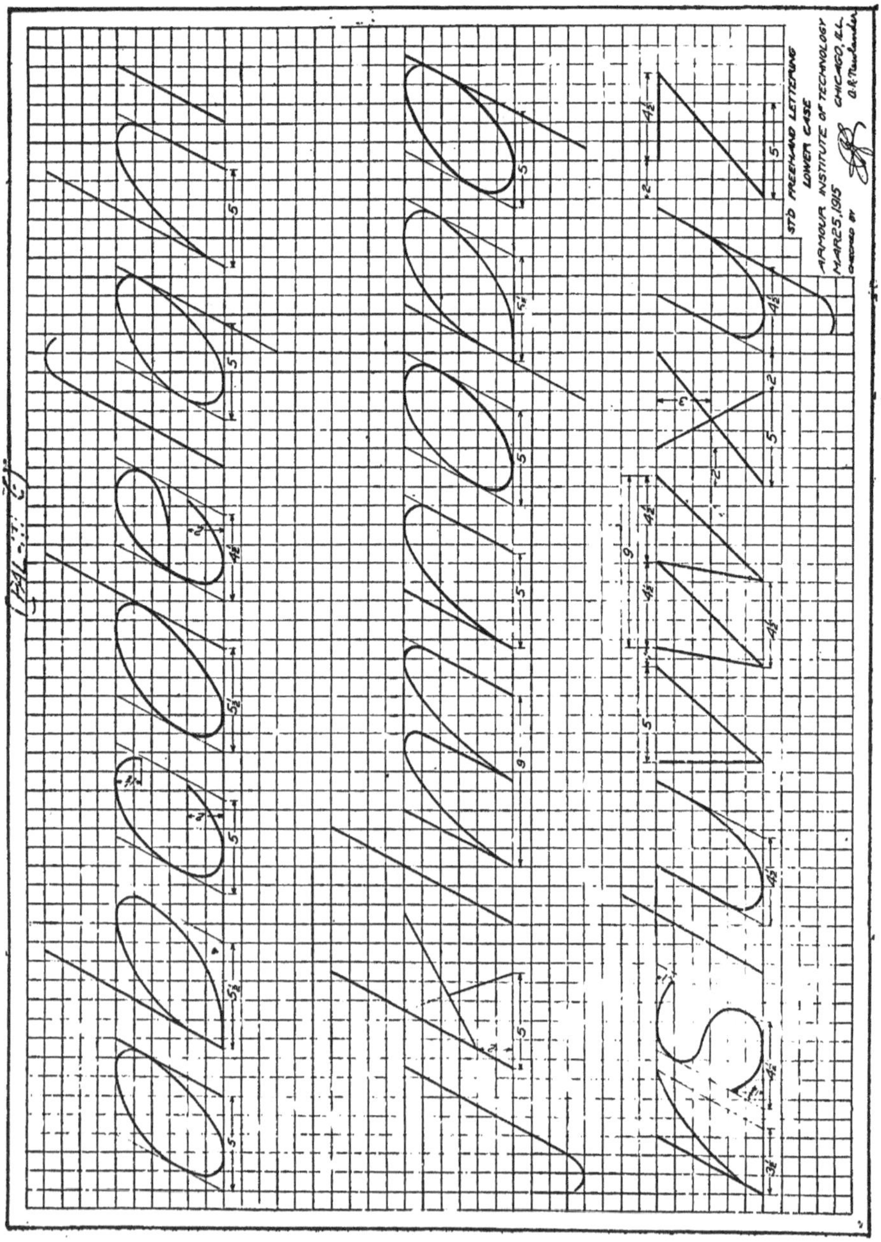

PLATE 9. 2 HOURS

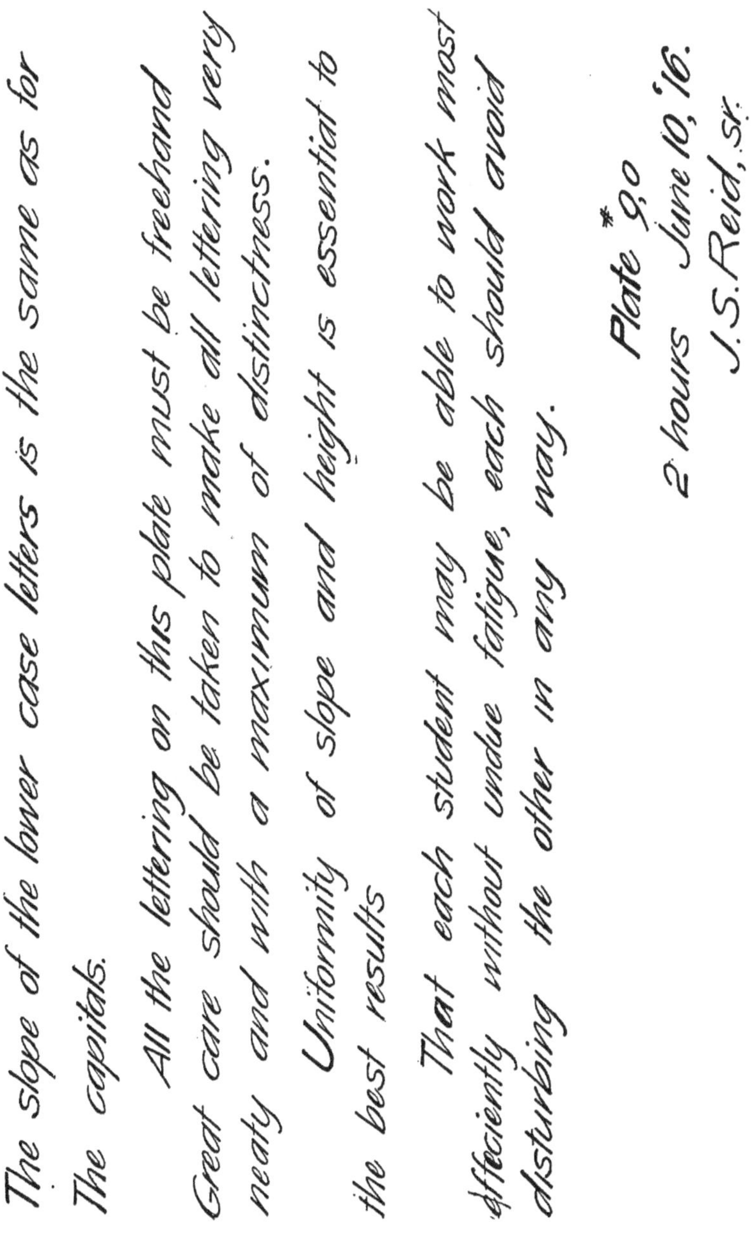

The slope of the lower case letters is the same as for
The capitals.

All the lettering on this plate must be freehand
Great care should be taken to make all lettering very
neatly and with a maximum of distinctness.

Uniformity of slope and height is essential to
the best results

That each student may be able to work most
effeciently without undue fatigue, each should avoid
disturbing the other in any way.

Plate *9,0
2 hours June 10, '16.
J.S.Reid, Sr.

FIG. 76.

PLATE 10. 4 HOURS

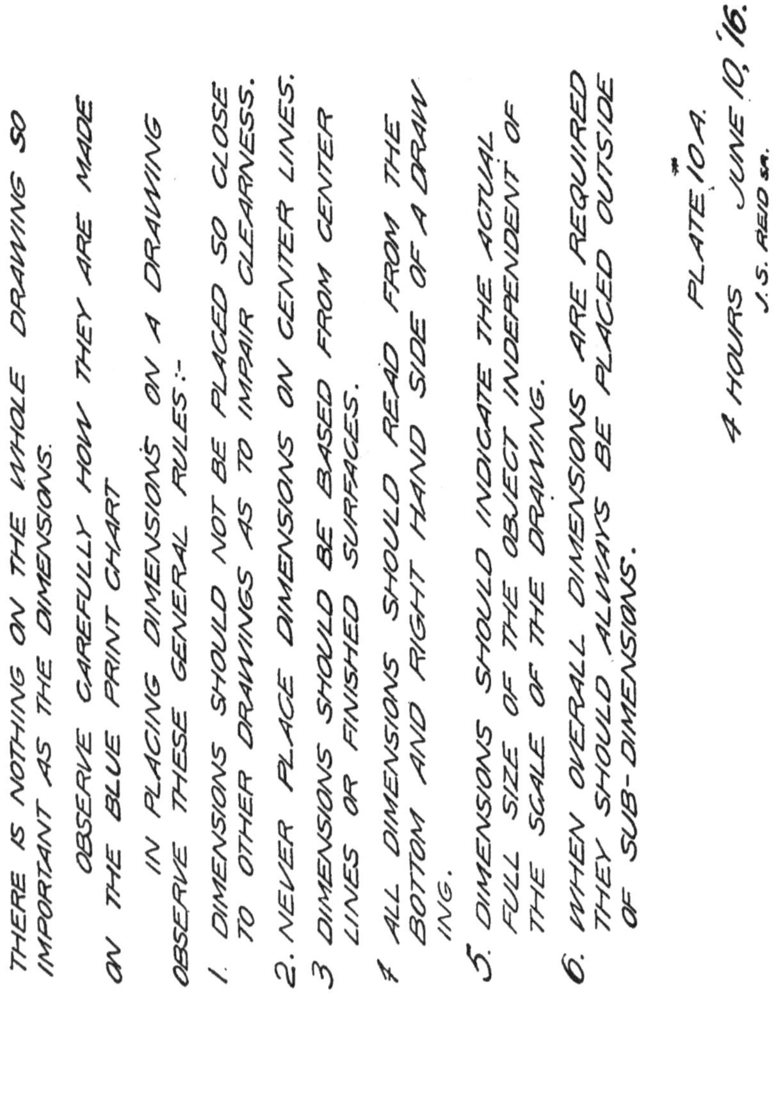

THERE IS NOTHING ON THE WHOLE DRAWING SO IMPORTANT AS THE DIMENSIONS.

OBSERVE CAREFULLY HOW THEY ARE MADE ON THE BLUE PRINT CHART

IN PLACING DIMENSIONS ON A DRAWING OBSERVE THESE GENERAL RULES:-

1. DIMENSIONS SHOULD NOT BE PLACED SO CLOSE TO OTHER DRAWINGS AS TO IMPAIR CLEARNESS.

2. NEVER PLACE DIMENSIONS ON CENTER LINES.

3 DIMENSIONS SHOULD BE BASED FROM CENTER LINES OR FINISHED SURFACES.

4 ALL DIMENSIONS SHOULD READ FROM THE BOTTOM AND RIGHT HAND SIDE OF A DRAWING.

5. DIMENSIONS SHOULD INDICATE THE ACTUAL FULL SIZE OF THE OBJECT INDEPENDENT OF THE SCALE OF THE DRAWING.

6. WHEN OVERALL DIMENSIONS ARE REQUIRED THEY SHOULD ALWAYS BE PLACED OUTSIDE OF SUB-DIMENSIONS.

PLATE 10A.
4 HOURS   JUNE 10, '16.
J. S. REID SR.

FIG. 77.

CHAPTER IV

ORTHOGRAPHIC PROJECTION

39. Orthographic Projection, sometimes called Descriptive Geometry and sometimes simply Projection, is one of the divisions of descriptive geometry; the other divisions are Spherical Projection, Isometric Drawing, Shades and Shadows, and Linear Perspective.

In this course we will take up only a sufficient number of the essential principles of Orthographic Projection, Isometric Drawing, Shades, Shadows and Shade Lines, to enable the student to make a correct commercial drawing of a machine or other object.

40. Orthographic Projection is the science and the art of representing objects on different planes at right angles to each other, by projecting lines from the point of sight through the principal points of the object perpendicular to the Planes of Projection.

There are usually three planes of projection used, viz., the **horizontal plane or H,** for short, the **vertical plane or V,** and the **profile plane or P.** Fig. 78.

Auxiliary planes are also used. These may be parallel, oblique or at right angles to *H* or *V* according to the conditions of the problem.

The *H* and *V* planes intersect one another at right angles in the line of intersection or ground line *GL* and form four dihedral angles.

The **first** angle is **above H** and in **front of V.**

The **second** angle is **above H** and **behind V.**

The **third** angle is **below H** and **behind V.**

The **fourth** angle is **below H** and in **front of V.**

41. The Profile Plane is perpendicular to both the V and H planes and is used in mechanical drawing to obtain end views of objects.

Fig. 78 is a pictorial view of V and H together with the right- and left-hand P planes. Fig. 79 is an orthographic end or profile view of these planes.

When viewing the planes of projection the observer should be stationed so that he is looking directly into the 1st and 4th angles. These angles then will be in *front* of V nearest

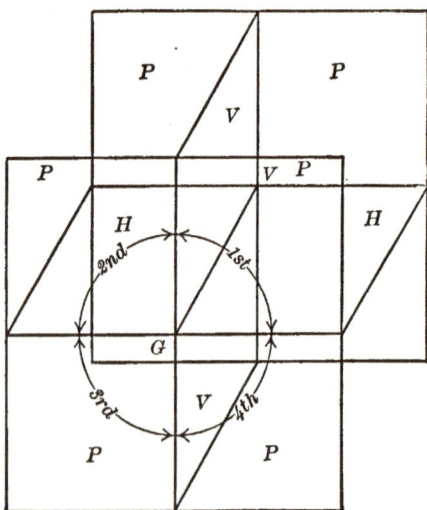

FIG. 78.

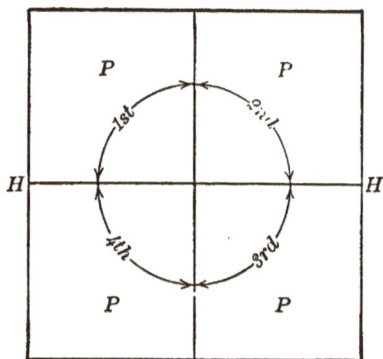

FIG. 79.

the observer while the 2d and 3d angles are behind V away from the observer.

According to Article 40 objects placed in space in any of these angles may be projected upon the V, H, and P planes by passing lines through their principal points perpendicular to the planes V, H, and P, respectively. These lines will pierce the planes of projection in points or traces. If these points are connected in proper sequence by lines, an orthographic projection of the object will be had on the three planes of projection.

But these planes are at right angles to one another and the three views are also at right angles to one another, while our drawing boards and drawing papers are flat; therefore, the planes containing the projections must be revolved into the flat to obtain a practical working drawing.

41. Revolution of the Planes. Let the horizontal plane represent the drawing board and paper lying in a horizontal position on the drawing table. Then if the V plane is revolved about the G.L. counter-clockwise until it coincides with H, the V projection of the object referred to above is revolved into the H plane or the plane of the paper either above or below the GL. If the object is in the 1st angle the V projection will show above G.L. after revolution and the H projection will remain below. The Profile Plane should be revolved so that the P projection will show in its proper quadrant after revolution according to the angle in which it is projected.

42. In order to still further explain the use of the planes of projection, with regard to objects placed in any angle, let us suppose a truncated pyramid surrounded by imaginary planes at right angles to each other, as shown by Fig. 80.

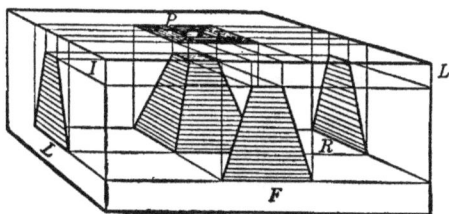

FIG. 80.

With a little attention it will easily be discerned that the pyramid is situated in the third dihedral angle, and that in addition to the V and H planes, we have passed two profile planes at right angles to the V and H planes, one at the right hand and one at the left.

When the pyramid is viewed orthographically through each of the surrounding planes, four separate views are had, exactly as shown by the projections on the opposite planes, viz., a Front View, Elevation or Vertical Projection at F, a

Right-hand View, Right-end Elevation or Right Profile Projection at R. A Left-hand View, Left-end Elevation, or Left Profile Projection at L.

A Top View, Plan or Horizontal Projection at P.

If we now consider the V plane and the right and left P planes to be revolved toward the beholder while looking down on H until they coincide, using the front intersecting lines as axes, the projections of the pyramid will be seen, as shown in Fig. 81, which, when the imaginary planes and pro-

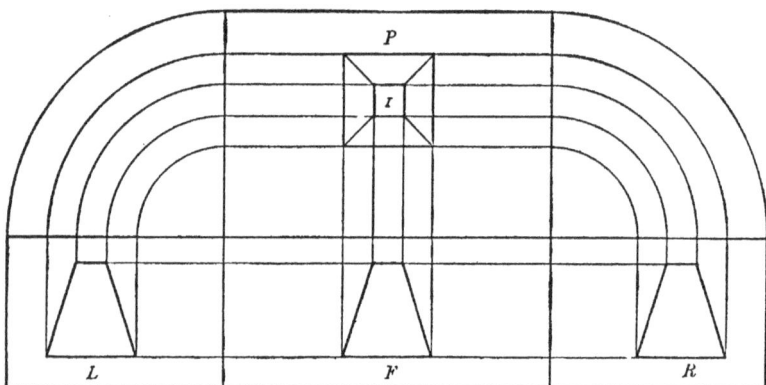

FIG. 81.

jecting lines have been removed, will be an orthographic projection or true drawing of the truncated pyramid.

43. To Illustrate Projections in the Different Angles. Let Fig. 82 be a pictorial view of the intersecting planes of projection V, H, and P, and B a rectangular block situated in space in the 1st angle. The observer in front looking toward the V plane can imagine parallel lines projected through the four corners of the nearest face of the block and prolonged toward V and perpendicular to it until these lines pierce V in four points called the traces of the lines. If these points are joined with straight lines parallel to each other, there will be formed on the V plane a V projection of the block in the 1st angle.

In a similar manner the observer viewing the block from above may project the top view of the block upon the H

plane determining the H projection of the block in the 1st angle seen at H.

The P projection may also be found in a similar manner by viewing the block from the right and projecting that view of the block upon P.

Authorities differ as to the proper way of viewing the object when projecting on the profile plane from the 1st angle. Some hold that in teaching descriptive geometry, the object should be viewed from the right in all angles when

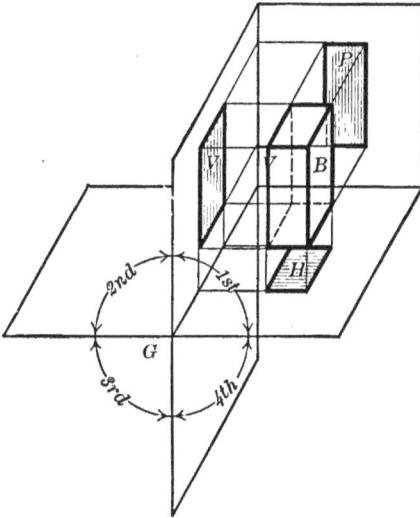

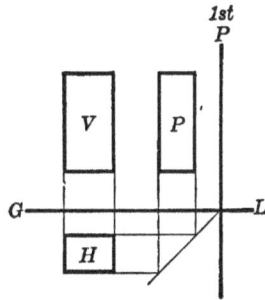

FIG. 82. FIG. 83.

projecting on the profile plane. It works all right in the 1st angle if the P line is placed far enough to the right of the V projection so that the profile projection when revolved (away from the observer) will not overlap the V projection. It places the right end view of the object directly beside the right end of the V projection, as seen in Fig. 83. But the plan or top view of the object remains *below* the V projection after the V plane has been revolved into the H plane or plane of the drawing board.

Others recommend that an object in the 1st angle should be viewed from the *left* when projecting on the P plane and from the right when viewing an object in the 3d angle.

What has already been said about projection in the 3d angle shows that it is the proper angle to use in projecting working drawings.

In Fig. 83 is shown the three views of the object on the flat after the P plane has been revolved into V and the V plane into H.

Fig. 84 shows the same block situated in space in the 3d angle, below H and behind V. The observer is located as

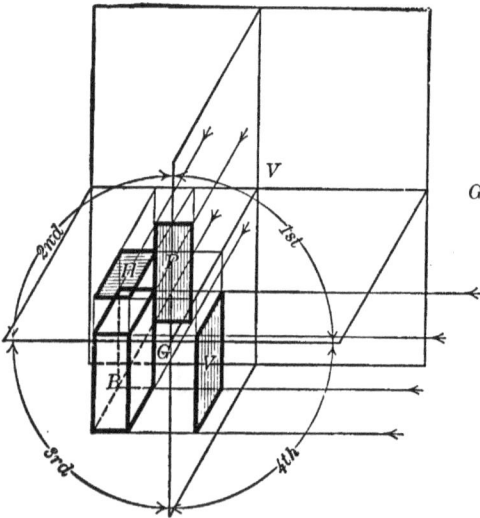

FIG. 84.

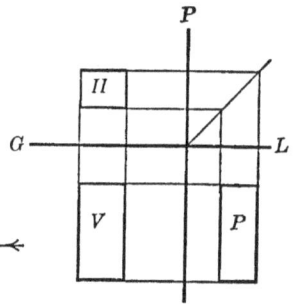

FIG. 85.

before in front of and looking toward the V plane. Parallel lines perpendicular to V are conceived to pass through the four corners of the front face of the block and where these lines pierce the V plane will be four points of the V projection of the block on the V plane. When these points are joined by parallel straight lines the V projection of the block will be complete. The plan or H projection is found in a similar manner looking from above toward the H plane. The P projection is found by looking toward the right-hand end of the block through the P plane, as shown. Fig. 85 shows the drawing of the block on the flat in the 3d angle after the planes have been revolved around the axes G.L. and *P* as explained in the projection in the 1st angle.

Projections in the 2d and 4th angles are seldom used except in solving problems in descriptive geometry. The method of projection, however, is the same as given for the 1st and 3d.

Commercial engineering drawings are made almost entirely in the 3d angle. The 3d angle is preferable for various reasons. Take, for example, the side view of a locomotive on the tract. In a drawing this would represent the vertical projection and in 3d angle projection the profile view of the front or smoke-box end would be placed directly beside itself to the right of the front and whereas in 1st angle projection the same front-end view would be projected across the vertical projection and placed at the left. See Fig. 86.

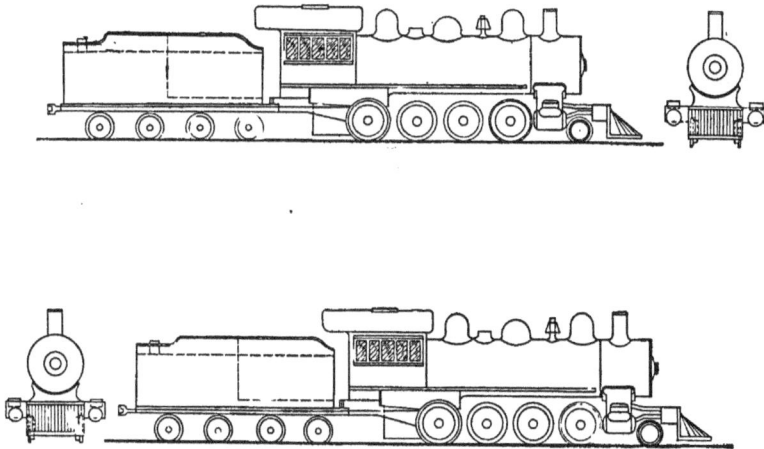

FIG. 86.

QUESTIONS ON CHAPTER IV

1. What is orthographic projection?. Art. 40.
2. What are the principal planes of projection?
3. What other planes are sometimes used?
4. What angles are formed by the *H* and *V* planes? Give names and how numbered.
5. Where is the *first* angle located with reference to the *H* and *V* planes?

6. Where is the second angle located with reference to the *H* and *V* planes?

7. Where is the third angle located with reference to the *H* and *V* planes?

8. Where is the fourth angle located with reference to the *H* and *V* planes?

9. Explain how objects in space in any of these angles are projected upon the planes of projection.

10. What is called the vertical projection?

11. What is the horizontal projection?

12. What is the profile projection?

13. What projection is called the *front view* or *elevation?*

14. What projection is called the *top* view or *plan?*

15. What projection is called the *end view* or *end elevation?*

16. How are the planes revolved to obtain the drawings on the flat?

17. How should the profile plane be revolved?

18. What angle does the projection lines make with the planes of projection?

19. What is the name of the line in which the *H* and *V* planes intersect?'

20. What angle is mostly used in projecting engineering drawings?

21. Explain why one angle is better than another in making working drawings.

CHAPTER V

The Representation of Points and Lines

44. Points. Having well in mind the representation of the Planes of Projection at right angles to each other forming the four dihedral angles as described in the previous chapter; it is easy for the student to conceive a point situated in space in any one of these angles and to project it upon the V and H planes by passing lines through the point perpendicular to H and V. The points in which these perpendicular lines pierce the V and H planes are the vertical and horizontal projections of the point in space, and when the V plane has been revolved away from the observer to coincide with H *the vertical and horizontal projections of the point will lie in a straight line perpendicular to* G.L.

45. Notation of Points. Capital letters will be used to designate points in space and lower case letters for the projections of points. The vertical projection to have a prime placed to the right and above the " point's " letter. See Fig. 87.

A represents the given point in space in the first angle, a' its vertical projection, a its horizontal projection and a'_p its profile projection.

46. Location of Points. Points are located with reference to their distance from the planes of projection. For example: $A(2''+1\frac{1}{2}''-1'')$ means that the point A is $2''$ from the left-hand border line, $1\frac{1}{2}''$ above G.L. and $1''$ below G.L. or in other words the point A is $1\frac{1}{2}''$ above H and $1''$ in front of V.

The first term in the parenthesis gives the distance from the left-hand border line measured along the G.L.

The second term is always the V projection whether plus

89

or minus. If *plus* it is laid up on a line perpendicular to
and *above* G.L. If *minus* it is laid off on a line perpendicular
to and' *below* G.L.

The third term is always the *H* projection whether plus
or minus. If plus it is located *above* and if minus, *below*, as
before.

A point located in the vertical plane will have its *H* projection in G.L. A point located in the horizontal plane will
have its *V* projection in G.L.

Example: $B(3,0+1)$ means that a point on G.L. $3''$ from
the left-hand border line is the *V* projection of the point *B*

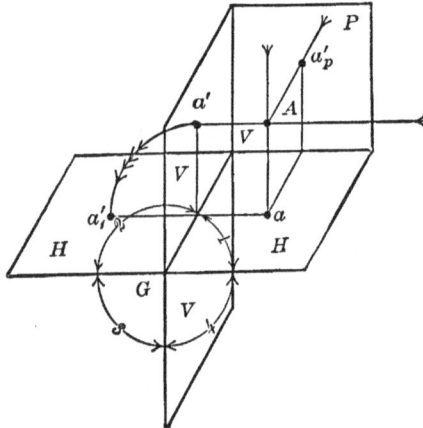

Fig. 87.

and on a line through that point perpendicular to G.L. and
$1''$ above it is located the *H* projection of the same point,
3d angle projection.

47. Projection of Points on the V, H and P Planes.
The projection of a point in the first angle is illustrated in
Fig. 87. The point *A* is shown in space above *H* and in
front of *V* and when projected on the *V* plane at *a'* and the
V plane revolved away from the observer until it coincides
with the *H* plane, then the point *A* is shown in vertical projection at a'_1, *above* the G.L. The *H* projection remains at *a'*
below the G.L.

The three projections of A on the flat are shown in Fig. 88. Notice that the P projection a_p' is at the left of the P line which is placed at a distance from the V projection suitable to enable the P projection to be revolved into the V plane without interfering with the V projection.

The projection of a point lying in H in front of V will have its V projection in the G.L. and its H projection on a line drawn through the V projection perpendicular to G.L. at a distance from G.L. equal to the distance of the point from the V plane. The P projection will be in G.L. to the left of P since the point is in front of V and viewed from the right. See Fig. 88.

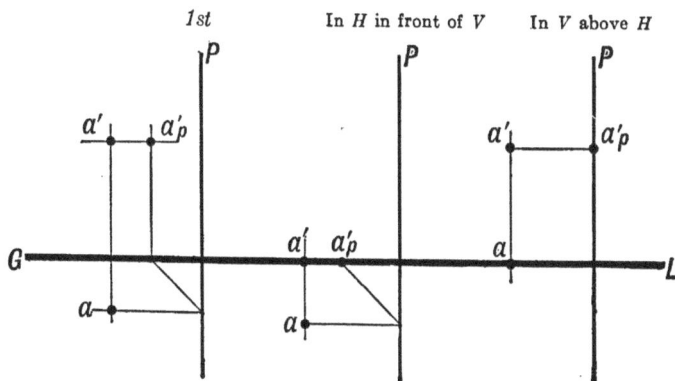

FIG. 88.

A point in V above H has its H projection in G.L. and its P projection in the P line since the P line is the end view of the V plane. See Fig. 88.

A point in space in the 3d angle at A, Fig. 89, viewed from the front through the V plane has its V projection below the G.L. and viewed from above its plan or H projection above G.L. Its P projection will revolve to the right of the P line below G.L. See Fig. 89.

A point in space in the 2d angle at A, Fig. 90, viewed from the front through V has its V projection above G.L. and viewed from above its H projection also falls above G.L.

Its P projection viewed from the right revolves to the *right* of the P line above G.L. into the 2d quadrant. See Fig. 91 for the projections on the flat.

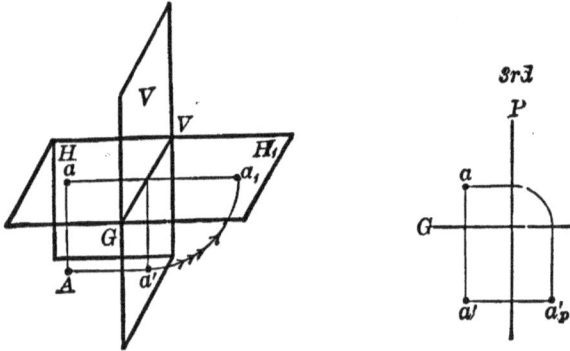

FIG. 89.

The projection of a point in space in the 4th angle is illustrated in Fig. 92. A is the point in space. When viewed from the front its V projection falls on the V plane below G.L. and its H projection viewed from above falls on the H plane below G.L.

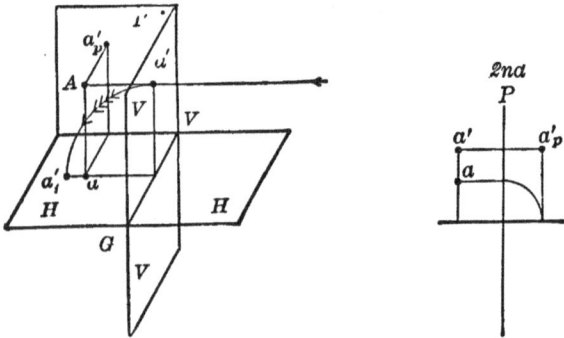

FIG. 90.

FIG. 91.

The P projection of a point in the 4th angle viewed from the right shows the point to the left of the P line in the 4th quadrant when revolved into the flat. See Fig. 91.

48. Plate 11. This plate will consist of the projection of

points, lines and planes.   Lay out the plate as shown in Fig. 93,
dividing the space inside the border line into 16 equal spaces.

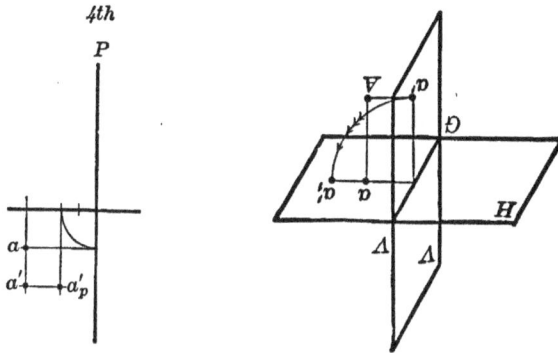

FIG. 92.

There will be 15 problems.   A title is to be carefully
lettered in the 16th space.   Use the standard title shown at
Fig. 35 and make the name of this plate " Projections."

All drawings are to be made first in fine, narrow lines
with the 6H pencil sharpened as described in Art. 4.

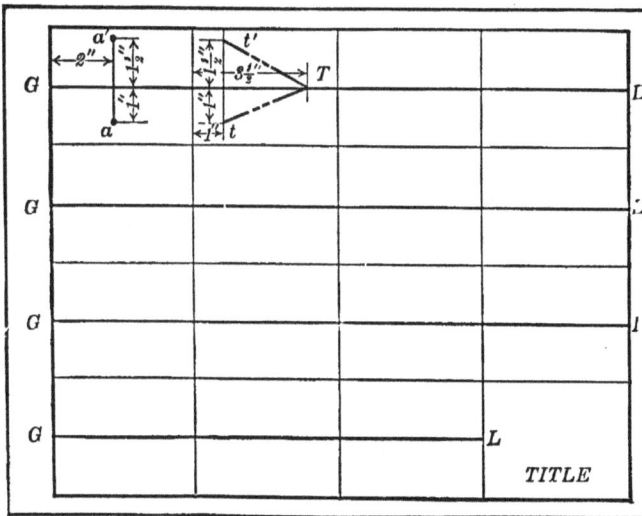

FIG. 93.

As soon as the projection of a point is located mark
it with its proper letter as directed in Art. 55, and designate

the angle in which it is projected in neat lettering above the projection of the point. See Fig. 88, page 91.

When all the problems in Plate 11 have been drawn in fine lines and approved (the problems should be submitted one by one for examination), all the given and required lines should then be strengthened with the 4H pencil, as described in Art. 35.

Projection lines must be made very narrow and unbroken.

Traces of planes will be shown by a long dash and two short dashes alternately. The short dashes about $\frac{1}{8}''$ long and $\frac{1}{16}''$ apart thus:

———— — — ———— — — ———————— Width about $\frac{1}{32}''$

PROB. 28. *Draw the V, H and P projections of the following points. Locate the P line to suit.*

$$A(\tfrac{1}{2}''+1''-\tfrac{1}{2}''),\ B(1\tfrac{3}{4}''-1''+\tfrac{1}{2}''),\ C(3''-\tfrac{1}{2}''-1'').$$

The projections of these points are to be drawn in the first division on the plate in the upper left-hand corner.

PROB. 29. *Draw the V, H and P projections of the following points:*

$$D(\tfrac{1}{2}''+0''-\tfrac{1}{2}''),\ E(1\tfrac{3}{4}''+\tfrac{1}{2}''-0'')\text{ and }F(3''+\tfrac{1}{2}''+1'').$$

49. Projection of Lines. The projections of two points of a straight line determine the projections of the line.

The projections of a line are the traces of the projecting planes of the line. The projecting planes are perpendicular to H and V, respectively, and intersect each other in the line itself and determine it in space.

In Fig 94, Ab' and Ba are two projecting planes at right angles to each other and perpendicular to V and H, respectively. ab is the H trace of the plane Ba and also the horizontal projection of the line AB. $a'b'$ is the trace of the plane $A \, b'$ and also the vertical projection of the line AB.

AB is the line of intersection between the two projecting planes and also the line itself in space.

When given the locations of two points as A and B of the line AB with reference to the planes of projection, proceed to draw their projections exactly as explained for the projection of points, and then draw a straight line from the V projection of A to the V projection of B as $a'b'$, Fig. 95, then $a'b'$ is the V projection of the line AB. Draw the H projections of the points A and B as a and b and join a and b with a straight line, then a, b is the H projection of the line AB. Fig. 95 shows the drawing on the flat.

The profile projection is found in the same manner as is used in finding the profile of points. See the profile pro-

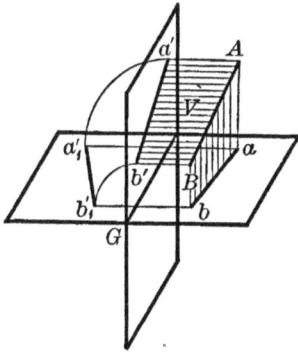

FIG. 94.

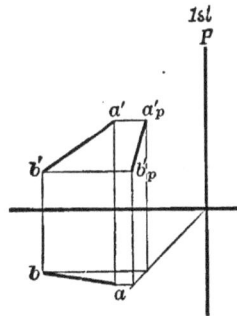

FIG. 95.

jection of point A in the 1st angle Fig. 88. A line is drawn through a parallel to G.L. to intersect P. The point when it intersects P is revolved into G.L. and a line erected at that point to meet a horizontal line through a' at a_p' the P projection of the point A in the 1st angle when viewed from the right. The method of drawing the P projection of the points A and B in Fig. 95 amounts to the same thing. The angle between the P line and G.L. below G.L. is bisected by a $45°$ line and horizontal lines drawn through a and b to meet it. At the points where these lines meet the bisector, erect perpendiculars to meet horizontals through a' and b' in the P projections of A and B in the 1st angle, as before, when viewed from the right.

50. If a line is parallel to H its V projection is parallel to G.L. Fig. 96a.

If a line is parallel to V its H projection is parallel to G.L. Fig. 96b.

If a line is oblique to H and V its projections are inclined to G.L. Fig. 96c.

If a line is perpendicular to H its projection on that plane will be a point and its V projection will be perpendicular to G.L. Fig. 96d.

If a line is oblique to H it will pierce H when prolonged.

The point where the line pierces H is called the piercing point, sometimes called the trace of the line. Fig. 96e.

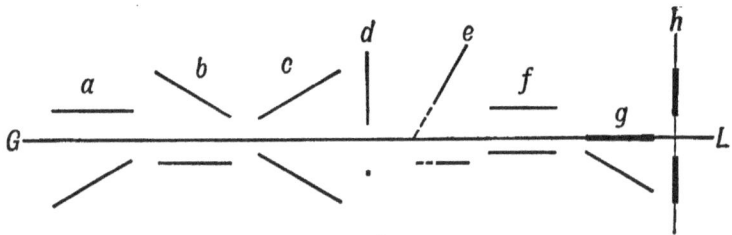

FIG. 96.

If a line is parallel to both H and V both of its projections are parallel to G.L. Fig. 96f.

If a line lies in H its H projection is the line itself and its V projection is in the G.L. Fig. 96g.

If a line lies in P its V projection will be in the V trace of the P plane and its H projection will lie in the H trace of the P plane. Fig. 96h.

It has been shown (Art. 49) that the projections of two points of a line determine the projections of the line. Two points are also necessary to determine a line in space as to its length and direction but one point and the direction of the line will also locate it in space.

51. If two lines intersect in space their projections will intersect in a *point* on a *line perpendicular to* G.L. Fig. 97.

Lines which do not intersect in space may intersect in projection but not in the same straight line perpendicular to G.L. Fig. 98.

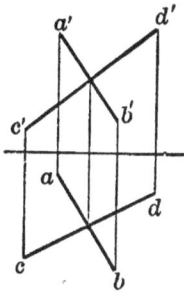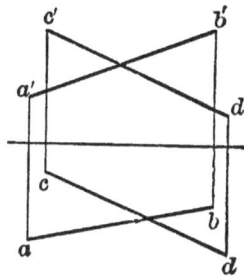

FIG. 97.   FIG. 98.

If two lines are parallel to each other in space, their projections will be parallel to each other on the same plane. Fig. 99.

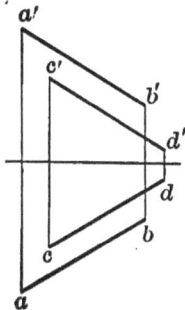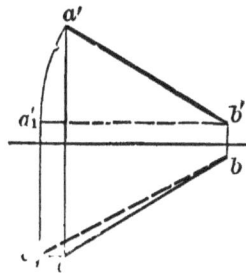

FIG. 99.   FIG. 100.

Lines oblique to V and H do not show their true lengths.

52. **To find the true length of a straight line oblique to both planes of projection.** a', b' and ab, Fig. 100, are the projection of a line in space oblique to V and H. Its true length may be found 1st by revolving parallel to V or H, 2d by revolving into H or V.

To revolve parallel to H. Take point b' as center and $b'a'$ as radius and revolve the line $a'b'$ until it is parallel to

G.L. at a'_1b'. During revolution the point a remains at the same distance from V, therefore, the projection of the arc $a'a'_1$, will be a straight line parallel to G.L. drawn through the point a to a_1 which is the H projection of a_1'. Join a_1b with a broken line. It is the true length of the line AB.

To revolve into H. Fig. 101.

At a and b erect perpendiculars to ab. At a lay off the distance aa_1 equal to the distance that a' is above G.L. At b lay off the distance bb_1 equal to the distance that b' is above G.L. At b lay off the distance bb_1 equal to the dis-

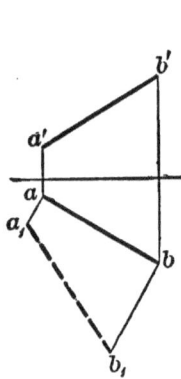

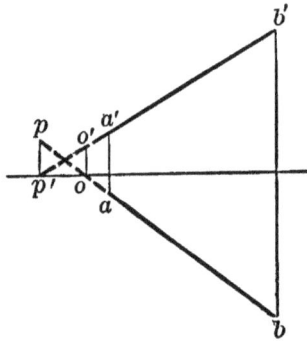

FIG. 101.   FIG. 102.

tance that b' is above G.L. Draw a_1b_1 in a broken line. It is the true length of the line AB.

53. To Find where a Straight Line Pierces the Principal Planes of Projection. A straight line in space oblique to the planes of projection will, if produced, pierce each of them in a point and since the piercing point lies in the plane its other projection will lie in the G.L.

For example: $a'b'$, Fig. 102, is the V projection of a straight line in space oblique to H. Extend $a'b'$ to meet G.L. at p^1 erect a perpendicular to meet the extended H projection of AB at the point p.

The point p is the H piercing point of the line AB. The V piercing point is found by erecting a perpendicular at the point O (where the H projection of AB meets G.L.) to meet the V projection.

PROB. 30. *Draw the V, H and P projections of the following lines. Locate the P line to suit.*

$$A(\tfrac{1}{4}''+\tfrac{7}{8}''-\tfrac{1}{4}'')\ B(1\tfrac{1}{4}''+\tfrac{1}{4}''-\tfrac{1}{2}'')$$

and

$$C(2\tfrac{1}{4}''-\tfrac{3}{8}''+\tfrac{7}{8}'')\ D(3\tfrac{1}{4}''-\tfrac{1}{4}''+\tfrac{1}{4}).$$

PROB. 31.

$$E(\tfrac{1}{4}''+\tfrac{1}{2}''+\tfrac{1}{4}'')F(1\tfrac{1}{4}+\tfrac{3}{4}+\tfrac{1}{4})$$

and

$$G(2\tfrac{1}{4}''-1''-\tfrac{1}{4}'')\ H(3\tfrac{1}{4}''-\tfrac{1}{2}''-\tfrac{3}{8}'')$$

PROB. 32.

$$K(\tfrac{1}{4}''+1''+\tfrac{1}{4}'')\ L(1\tfrac{1}{4}''-1''-\tfrac{1}{4}'')$$

and

$$M(2\tfrac{3}{4}''+\tfrac{1}{2}''+o'')\ N(3\tfrac{3}{4}''+\tfrac{1}{2}''+o'').$$

PROB. 33. *Draw the H projections of the two lines*

$$A(\tfrac{1}{4}''+1''-X'')\ B(1\tfrac{1}{4}''+\tfrac{1}{4}''-X)$$

and

$$C(\tfrac{1}{4}''+\tfrac{1}{4}-X)D\ (1\tfrac{1}{4}''+1''-X''),$$

intersecting in space, see Art. 51.

PROB. 34. *Assume the H projections of two lines parallel to each other in space in the first angle and draw the V projections.*

PROB. 35. *Find the V and H piercing points of the lines*

$$A(\tfrac{1}{4}''-1''+\tfrac{3}{4}'')\ B\ (1\tfrac{1}{4}''-\tfrac{1}{4}''+\tfrac{3}{8}'')$$

and

$$C(3''+\tfrac{1}{2}-1'')\ D(4''+\tfrac{1}{2}-\tfrac{1}{4}'').$$

CHAPTER VI

Representation of Planes

54. Representation of Planes. Many other planes besides H V and P may be used in projection. Such planes are shown by their traces on the principal planes of projection.

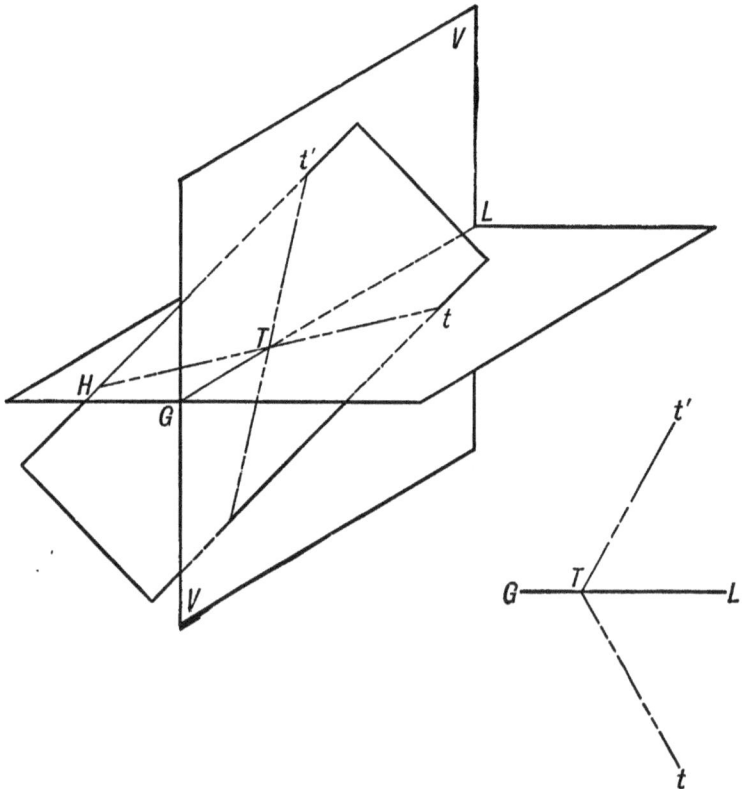

Fig. 103.

Fig. 103 gives a pictorial view of an oblique plane T intersecting H and V. Tt' is the V trace of the plane or

in other words it is the line in which the oblique plane penetrates or intersects V. Tt is its H trace or the line in
which the oblique plane T intersects the H plane. The
orthographic drawing of the traces of the plane is shown at
the right in Fig. 103.

Given two traces of a plane the plane is completely
determined.

55. Location of Planes. Since planes are determined by
their traces it is sufficient to give the location of the traces
with reference to H and V to completely determine the plane.

The traces of all oblique planes must meet at G.L. This
point is noted with a capital letter as T in Fig. 103, and the
plane is known as the plane T.

Traces of planes are located in a similar manner to that
used in locating points, for example:

$$T(1''+1\tfrac{1}{2}'')3(1''-2'').$$

means that the V trace of the plane T has its point t'
located $1''$ from the left-hand border line and $1\tfrac{1}{2}''$ above
G.L. and its point t $1''$ from the left-hand border line and
$2''$ below G.L. while the point of intersection of the two
traces on G.L. is $3''$ from the left-hand border line. See
Fig. 93, Art. 48.

Note that in the first parenthesis the 1st figure is the distance from the left border line to the projection line of a
point on the V trace. The 2d term, viz., $''+1\tfrac{1}{2}''$ (or whatever the value may be) is always the V projection of a point
on the V trace whether plus or minus.

The figure between the parenthesis is always the distance of the meeting point of the two traces on G.L. from
the border line.

In the second parenthesis the first figure is the distance
from the left-hand border line to the projection line of a
point on its H trace as t. The second term, viz., $''-2''$ is
the distance that the point t on the H trace is from G.L.

56. Planes. Planes may be determined in space by *three
points not in the same straight line*, by *two intersecting lines*,

two parallel lines, or by *one point and a straight line*. Planes
extend indefinitely so that they intersect one or both of
the planes of projection.

The lines in which a plane intersects the planes of projection are called the traces of the plane. These trace lines
may be made of long dashes and two short dashes alternately,
as shown in Fig. 104.

*To find the traces of a plane given by three points A, B
and C.* Fig. 104.

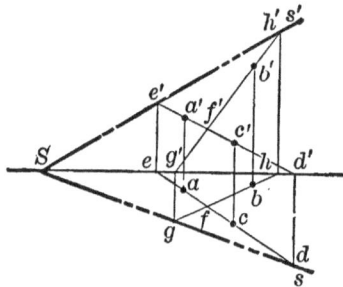

FIG. 104.

Draw through any two of the points a straight line as AC
and produce it to pierce H at D and V at E. d and e' are
points in the horizontal and vertical traces respectively.
Draw through the remaining point B a straight line to intersect DE in any point F and produce BF to pierce H and
V at G and H. g is a second point in the horizontal and h'
a second point in the vertical trace of the plane S.

PROB. 36. *Draw the traces of the plane given by three points*
$A(1\frac{1}{2}'' + \frac{3}{4} - \frac{3}{8}'') B(2\frac{3}{8}'' - \frac{3}{4}'' - 1'')$ and $C(2\frac{1}{2}'' + 1\frac{1}{8}'' + \frac{3}{8})$.

**57. To Pass a Plane through Two Intersecting Straight
Lines.** Find the V and H piercing points of the lines. They
will be points in the traces of the plane. In case the lines
do not pierce the planes of projection within the limits of
the drawing, connect a point on each with a straight line and
find its V and H piercing points. Another piercing point
found in a similar way will determine the plane.

The same method will apply in passing a plane through

two parallel straight lines or through a point and a straight line.

PROB. 37. *Assume the V and H projections of two lines intersecting each other in space and draw the traces of the plane containing them.*

58. Given the traces of two planes to find the line of intersection between them.

The point of intersection A, Fig. 105, of the two vertical traces is one point in the required line. The point of intersection B of the horizontal traces is another. Join these two points with a straight line. It is the required line of intersection.

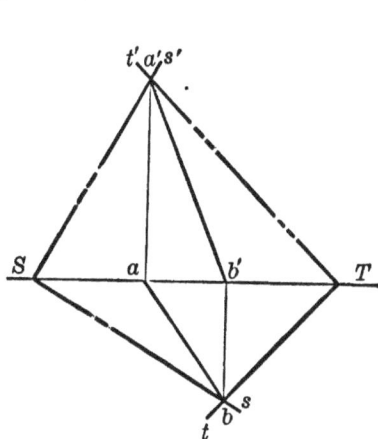

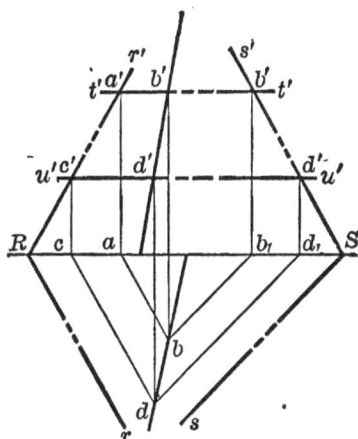

FIG. 105.   FIG. 106.

59. To Find the Line of Intersection of Two Planes whose Traces do not Intersect within the Specified Limits. Pass two planes parallel to H. These auxiliary planes cut from the given planes straight lines parallel to their horizontal traces.

T and U, Fig. 106, are the traces of the auxiliary planes parallel to H. ab and $a'b'$, cd and $c'd'$ are the projections of the lines cut from the planes R and S. Draw a line through db and $d'b'$. These lines are the horizontal and vertical projections of the line of intersection between the two planes R and S.

PROB. 37. *Find the line of intersection between the planes*
$T(2''+1\frac{3}{8}'')\frac{3}{4}(2\frac{1}{2}''-1)$ and $S(2\frac{1}{4}''+1\frac{1}{2}'')3\frac{3}{4}''(2\frac{1}{2}''-1\frac{1}{8}'')$.

PROB. 38. *Find the line of intersection between the following
planes whose traces do not intersect:*

$U(2''+1\frac{1}{8}'')1(1\frac{7}{8}''-1$ and $V(3+1\frac{1}{8})4(2\frac{3}{4}''-1)$.

**60. To Project a Given Straight Line on a Given Oblique
Plane.** T is the given plane and AB the given straight line,
Fig. 107.

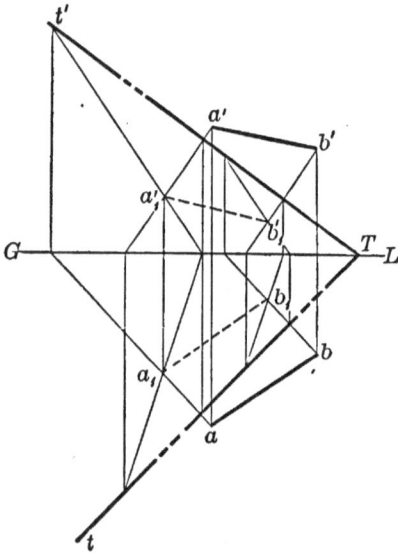
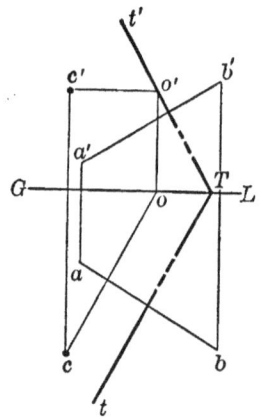

FIG. 107.   FIG. 108.

Assume two points on the line, say, A and B, and through
them draw lines perpendicular to the given plane. They will
be perpendicular to the traces of the given plane. Find their
piercing points and join them with a straight line, as shown.
$a_1b_1a'_1b'_1$ are the projections of the line on the plane.

PROB. 39. *Project the line $A(\frac{7}{8}+\frac{1}{2}-\frac{5}{8})$, $B(1\frac{1}{2}+1-1\frac{1}{4})$ on the
plane $T(2\frac{1}{2}+1(1)2\frac{1}{4}-1\frac{1}{2})$.*

**61. To Pass a Plane through a Given Point Perpendicular
to a Given Straight Line.** Through the point C, Fig. 108,

draw a line perpendicular to *ab* and parallel to *H* and find
its piercing point *O*. Through *O* draw the traces of the required plane perpendicular to the line *AB*.

PROB. 40. *Pass a plane through the point* $A(1-1\frac{1}{2}+1$ *per-
pendicular to the line* $C(1\frac{1}{2}+1\frac{1}{4}-\frac{1}{4})$, $D(3+\frac{1}{2}-1\frac{1}{4})$.

**62. To Pass a Plane Parallel to a Given Plane at a Given
Distance from it.** *T*, Fig. 109, is the given plane. Pass a
plane *R* perpendicular to *T* and revolve it into *H*, thus:
Erect a perpendicular at *R* and lay off Rr'_1 equal to Rr'.

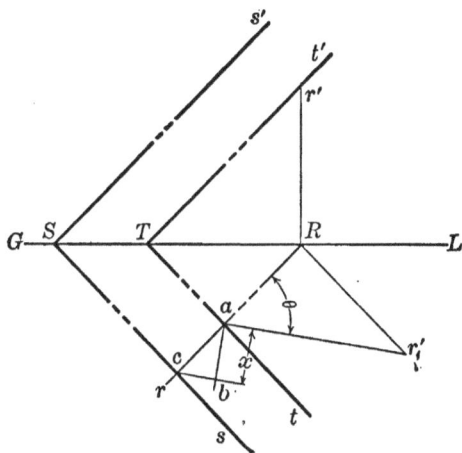

FIG. 109.

Join *a* and r'_1. It is a line cut from the plane *T* revolved
into *H*. Perpendicular to this line lay off the given distance
ab equal to *x*. Through *b* draw a line parallel to ar_1' cutting
rR in the point *c*. Through *c* draw the *H* trace of the
required plan *S* parallel to *Tt*. Draw the *V* trace Ss' parallel to Tt'. Angle θ is the angle which the plane *T* makes
with *H*.

PROB. 41. *Draw the traces of a plane parallel to plane*
$T(\frac{1}{2}+1\frac{1}{2})3(\frac{1}{2}-1\frac{1}{2})$ *and* $\frac{1}{2}''$ *from it. Measure the angle the
required plane makes with V.*

**63. Given the Traces of Two Oblique Planes to Measure
the Angle between Them.**

1st. Find the line of intersection between the given planes S and T, Fig. 110.

2d. Pass a vertical projecting plane perpendicular to the line of intersection $r'r'$; is its V trace. This auxiliary plane will cut from the given planes two straight lines which form a triangle with $a'b'$ as its base and the point c' on the line of intersection its vertex.

3d. Revolve the point c' around the line $r'r'$ into V at c_1' join c_1' and the points $a'b'$ with straight lines.

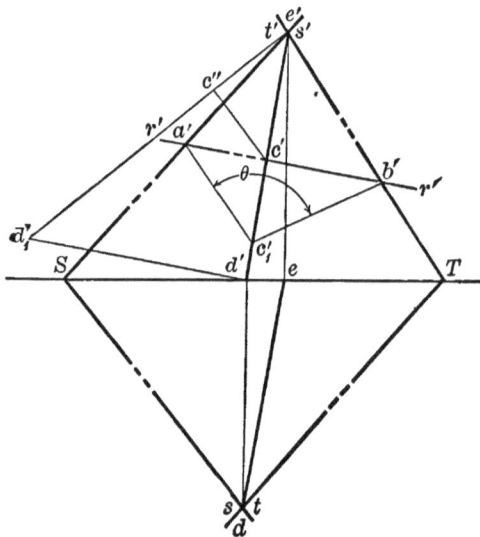

FIG. 110.

The angle θ contained between them is the angle required.

To obtain the distance c_1' from c', revolve the line of intersection into V, by erecting a perpendicular at d' and laying off the distance $d'd_1'$, equal to $d'd$. Draw $d_1'e'$. It is the revolved position of the line of intersection. Draw $c'c''$ perpendicular to $d_1'e'$. It is the distance required.

PROB. 42. *Find the angle between the planes* $S(1\frac{1}{4}+1\frac{1}{2})\frac{1}{2}$ $(1\frac{7}{8}-1\frac{1}{2})$, *and* $T(2+1\frac{1}{2})3(2\frac{1}{4}-1\frac{1}{2})$. *Also find the angle which* T *makes with* H.

64. To Find the Shortest Distance between Two Given Straight Lines.

Let AB and CD, Fig. 111, be the two given straight lines.

1st. Pass a plane T through CD parallel to AB.

2d. Project the line AB on this plane (Art. 60) thus: Through any point on AB, as H, draw a line perpendicular to the traces of T and find its piercing point on plane T at E. This is one point in the projection of AB on plane T.

3d. Draw through E the line EF parallel to AB. It will intersect CD, which is also lying in plane T in the point F.

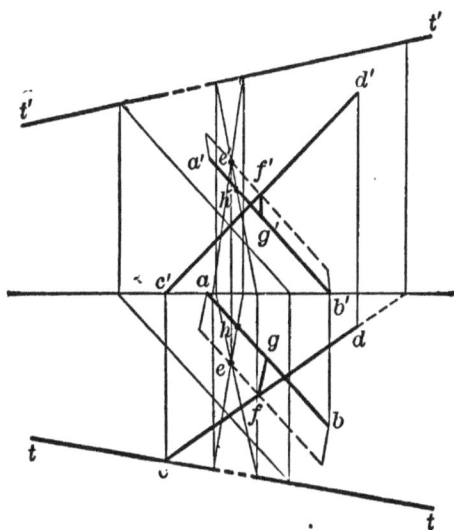

FIG. 111.

4th. Draw through F the line FG perpendicular to the plane T. It is perpendicular to both lines and is, therefore, the shortest distance between them. The true length of FG may be found by Art 52.

PROB. 42. Measure the shortest distance between the straight lines $A(\frac{3}{4}+1\frac{1}{2}-0)$, $B(2\frac{3}{4}+0-1\frac{1}{2})$, and $C(1\frac{1}{4}+0-1)$, $D(2\frac{1}{2}+1\frac{1}{4}-0)$.

65. To Draw the Projections of a Straight Line 2″ long Making an Angle 45° with H and an Angle of 30° with V.

1st. Assume the point a in H, Fig. 112, and draw through it line ab 2'' long, making an angle of 30° with G.L. $a'b'_2$ is its V projection.

2d. At the point a' draw $a'b'_1$ 2'' long, making an angle of 45° with G.L. ab_1 is its H projection.

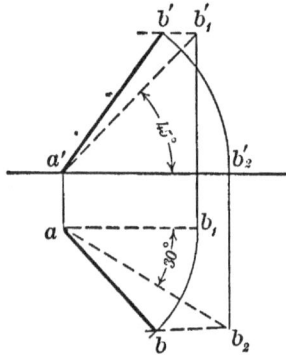

FIG. 112.

3d. Using a and a' as centers and ab_1 and $a'b'_2$ as radii describe arcs cutting lines parallel to G.L. through b_2 and b_1' in the points b and b'.

4th. Join a and b and a' and b' with straight lines. They are the projections required.

PROB. 43. *Draw the projections of a line making an angle of 30° with H and an angle of 45° with V.*

CHAPTER VII

ORTHOGRAPHIC PROJECTION APPLIED

THE problems in this chapter give an opportunity for the student to apply the principles of orthographic projection to the making of practical working drawings.

66. PLATE 12. This plate will consist of eight complete projections of solids of various forms and familiar outlines. Divide the plate into 9 equal spaces, as shown in Fig. 113.

Fig. 114 shows an isometric drawing of a wedge in which there are eight principal points which, when properly projected according to the principles given in Art. 47 and joined in correct order with straight lines will produce a complete mechanical drawing in three views.

PROB. 44. *In the upper left-hand space on the plate draw the V, H and P projections of the wedge shown in Fig. 114.*

This problem, like all the others in Plate 12, is to be drawn very carefully and accurately with fine, narrow lines, using the 6H pencil properly sharpened. Each problem, when completed should be submitted for approval and when correct in every particular it should be checked by the instructor and directions given for the next problem.

PROB. 45. *The same wedge used in Prob. 44 is to be drawn in space 2, but in a reversed position, as shown in Fig. 113. Draw the V, H and P projections.*

It will be noticed in the plate layout, Fig. 113, that the P projection in Probs. 44 and 45 is revolved by two different methods. The first by arcs of circles, centered at the point of intersection of P and G. L. the second by straight lines drawn from the points to be revolved to the bisector of the angle. The bisector is a line drawn through the same point of intersection

and making the angle of 45° with G. L. The latter method is
the most convenient since the revolution can be accomplished
by drawing straight lines with the T-square and triangle.

PLATE 12

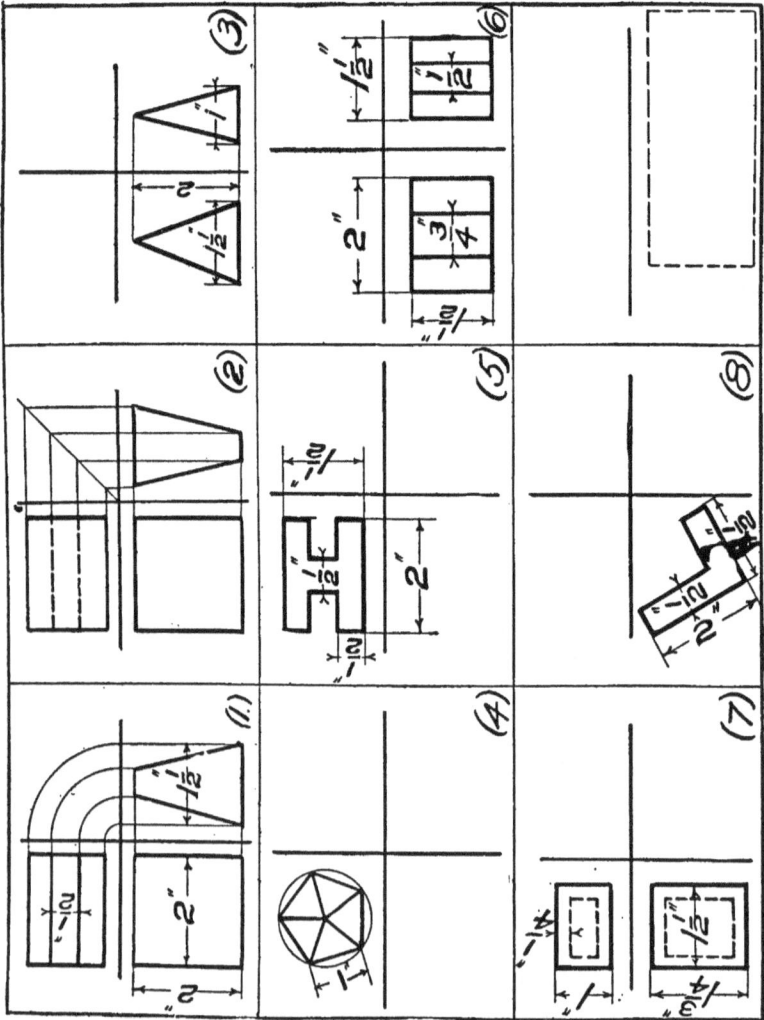

FIG. 113.

PROB. 47. *Given the V and P projections of the rectangular
pyramid illustrated in Fig. 115. Find and draw the H projection.*

For the definition of a rectangular pyramid, see **Art.** 31, No. *O* 74.

PROB. 48. *Given the H projection of a pentagonal pyramid draw the V and P projection.* Fig. 116.

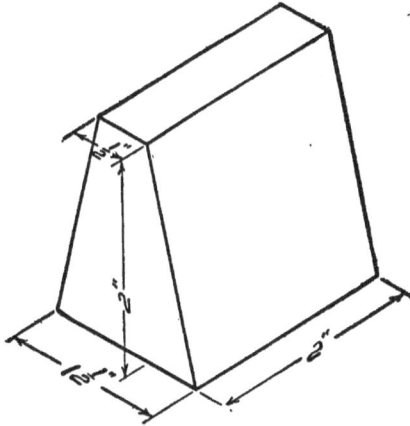

FIG. 114.

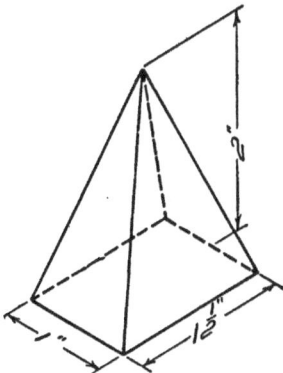

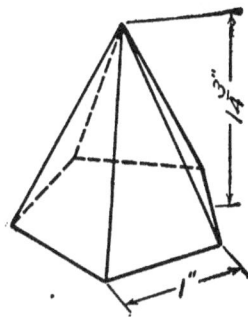

FIG. 115.   FIG. 116.

Draw the pentagon by the method shown in Prob. 10, Fig. 50, page 56.

PROB. 49. *Given the plan of an H-shaped solid, draw the* **elevation** *and* **right-end view.** Fig. 117.

PROB. 50. *Given the* **front elevation** *and* **end elevation** *of a cross-shaped solid, draw the* **plan.** Fig. 118.

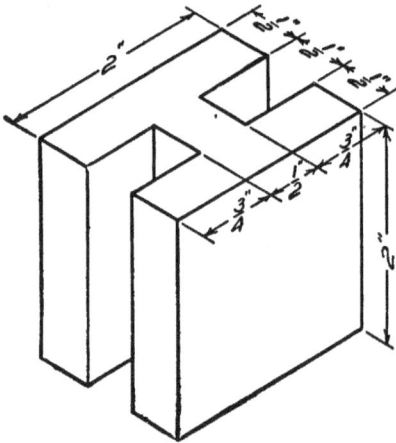
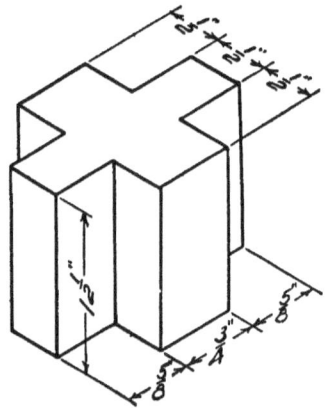

FIG. 117. FIG. 118.

PROB. 51. *Given the* **elevation** *and* **plan** *of the rectangular box shown in Fig.* 119. *Draw the* **right-end** elevation.

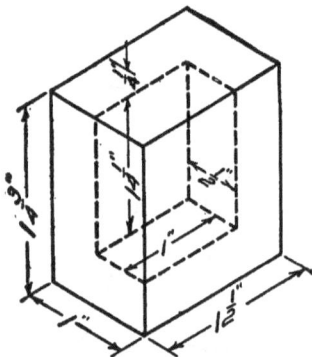
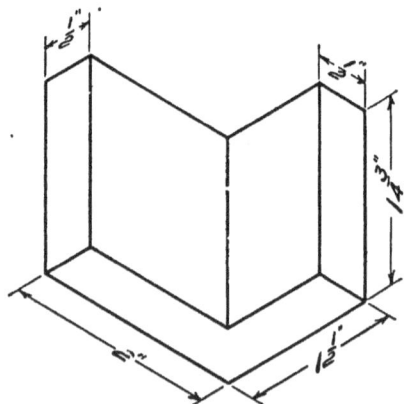

WALLS ALL ⅛ THICK

FIG. 119. FIG. 120.

PROB. 52. *Given the V* **projection** *of the L-shaped solid shown in Fig.* 120. *Draw the* **H** *and* **P** projections.

When all the problems in Plate 12 have been drawn correctly in fine lines and checked, the plate should be cleaned with art gum and the object lines of all the drawings strengthened with the 4H pencil·properly sharpened.

2d. The dimension lines should next be drawn, very narrow with the 6H pencil.

3d. Put on all dimension figures and arrow points, first the left-hand arrow point then the dimension and sign of inches and then the right-hand arrow point.

Instead of the problem numbers 1,˙2, 3, etc., as given in Fig. 113, use the numbers of problems given in the text as 44, 45, etc., thus " Prob. 44." Letters $\frac{3}{32}''$ and figures $\frac{3}{16}''$ high.

4th. Letter the title in space reserved in lower right-hand corner observing the directions given for title in Plate 2, Art. 35. The main title to be " Ortho. Projection."

5th. Ink border line, problem numbers, and title.

67. PLATE 13. *Draw the rectangular prism, Fig 121, in eight different positions similar to the position of the square pyramid shown in Fig. 122,˙ and according to the directions given below.*

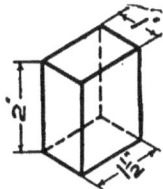

FIG. 121.

PROB. 53. *Given the elevation and plan of the prism, Fig. 121 draw the end view.*

Prob. 54. Given the same prism of problem 53 when the plan has been rotated to the left through an angle of 15°. Project the front and end elevations.

PROB. 55. *Given the front elevation of the figure obtained in problem 54 when revolved to the left through an angle of 15°. Draw the plan and end elevation.*

PROB. 56. *Given the front elevation of problem 53 when revolved through an angle of 30° to the right. Draw the plan and end view.*

PROB. 57. *Given the end elevation of the pyramid obtained in Prob. 54 when revolved to the right through an angle of 15°. Project the front elevation and plan.*

PROB. 58. *Given the end view of the pyramid obtained in Problem 55 when revolved to the left through an angle of 45°. Draw the front elevation and plan.*

PROB. 59. *Given the end view of the pyramid obtained in Problem 56 when revolved through an angle of 30° to the left. Draw the elevation and plan.*

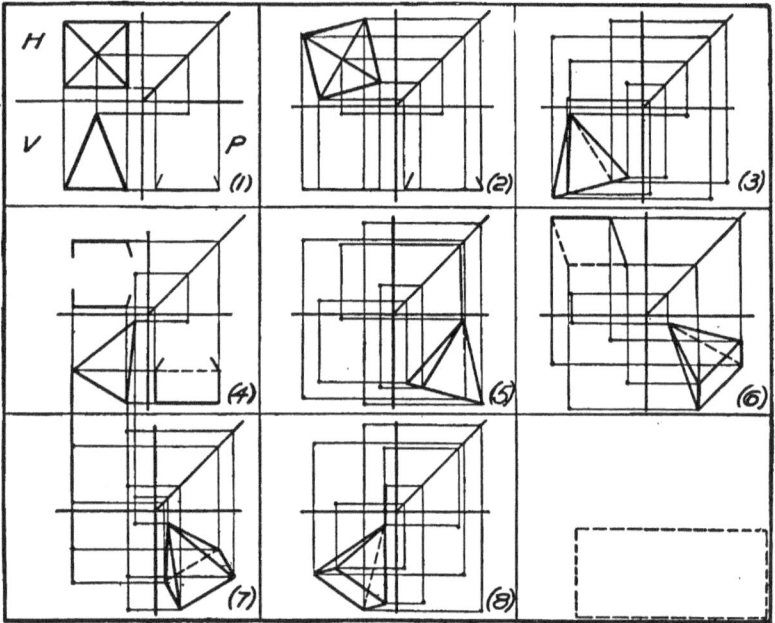

FIG. 122.

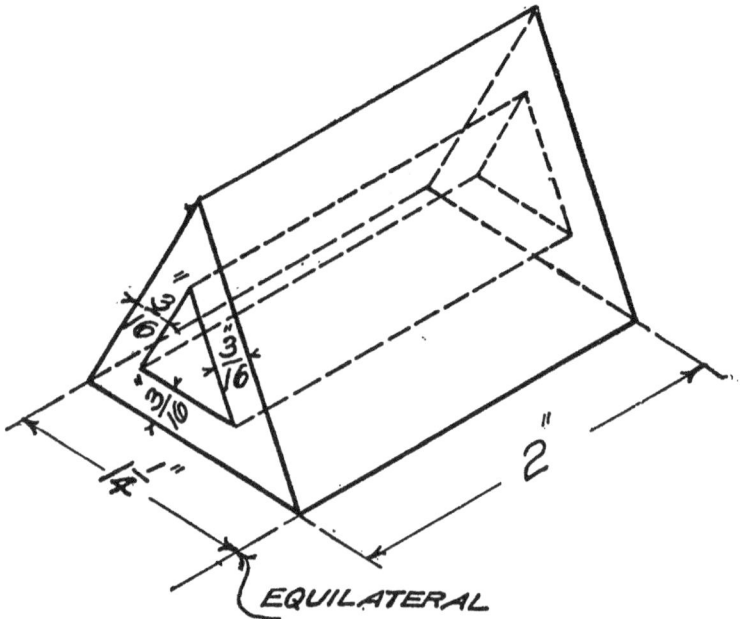

EQUILATERAL

FIG. 123.

PROB. 60. *Given the front elevation obtained in Problem* 57 *when revolved* 30° *to the right. Draw plan and end view.*

67. Auxiliary Planes. It is sometimes necessary to show a detail of a machine or a section on a plane oblique to the principal planes of projection. Such planes are called auxiliary planes.

PLATE 14. This plate will consist of five problems to illustrate the use of the auxiliary plane.

PROB. 61. *Given the elevation and plan of the hollow wedge-shaped piece shown in Fig.* 123.

1st. Draw an end view of the wedge at the left of the position of the *V* projection, shown in Fig. 124, taking the dimensions from Fig. 123, and arrange the views in a convenient manner in the upper left-hand corner of the plate.

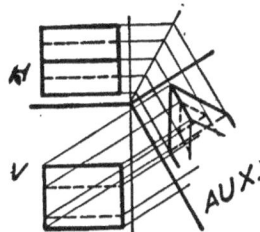

FIG. 124.

2d. Draw the auxiliary *P* line making 30° with *P* and draw an auxiliary ground line at right angles to the auxiliary *P* line, as shown in Fig. 124.

3d. Draw fine, narrow lines from principal points in elevation perpendicular to auxiliary *P* line.

4th. Draw perpendicular lines from corresponding points in the plane to the *P* line extending them to intersect the bisector of the angle between *P* and the auxiliary G.L.

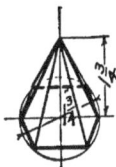

FIG. 125.

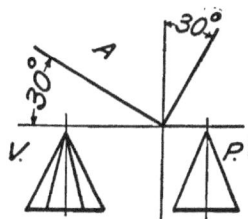

FIG. 125a.

5th. From the points of intersection on the bisector, draw lines perpendicular to the auxiliary G.L. to meet the corresponding lines from the elevation and complete the auxiliary view similar to the partly drawn view shown in Fig. 124.

PROB. 62. *Arrange the front and end view of the hexagonal pyramid given in Fig. 125 to obtain an auxiliary view looking in a*

FIG. 126.

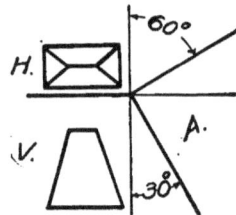

FIG. 127.

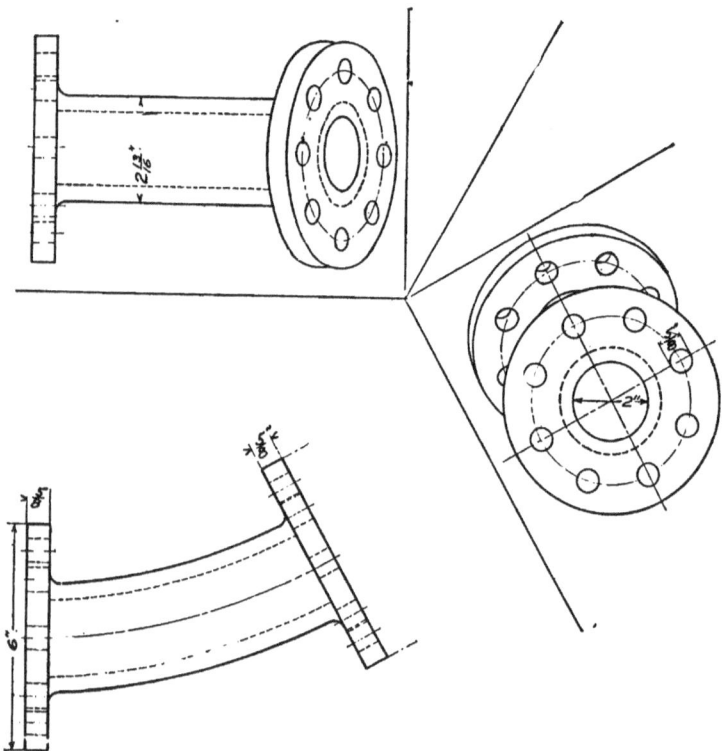

FIG. 128.

direction perpendicular to the auxiliary G.L. which is to make an angle of 30° with H, as shown in Fig. 125a.

PROB. 63. *Given the elevation and plan of the wedge shown in Fig.* 126. *Draw an auxiliary view looking in the direction as indicated in Fig.* 127.

PROB. 64. *Given the plan and elevation of the pipe elbow shown in Fig.* 128. *Draw an auxiliary view perpendicular to the face of the inclined flange. This view will give the true size of the flange and the holes through it.*

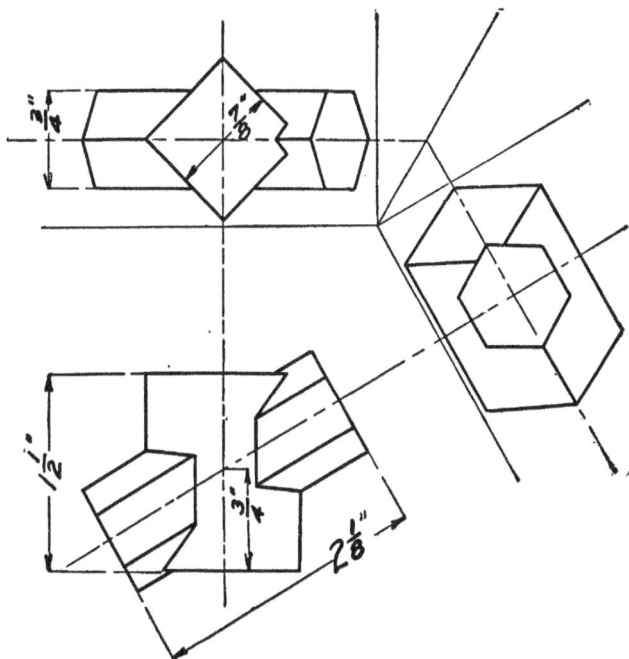

FIG. 129.

PBOB. 65. *Given the elevation and plan of the square prism penetrated by the hexagon prism as shown in Fig.* 129. *Draw an auxiliary view perpendicular to the hexagon prism.*

In laying out the drawing of the elevation and plane, draw all the center lines first.

2d. Draw the inscribed circle of the square prism in the plan and project the sides to the front view.

3d. On the center line of the hexagonal prism construct a hexagon whose inscribed circle shall be equal to $1\frac{5}{8}''$.

4th. From the six corners of the hexagon draw the sides of the hexagonal prism in the elevation.

5th. Draw the auxiliary view as required and determine the V and H projections of the lines of intersection between the hexagonal and square prism.

Main title. " Projection on Auxiliary Plane." This plate is to be finished in pencil. Ink title, problem numbers and border line.

SECTIONS

68. Intersections and Developments.

SECTIONS. When an object is cut by a plane the surface seen when a part is removed is called a *section*.

When it is difficult to show the interior construction of an object by invisible lines, it is usual to pass a cutting plane through that part of the object desired to be shown and when the part to the right or left of the cutting plane is removed there is obtained what is called a sectional view.

INTERSECTION. The piercing point of a line where it penetrates a plane is the point of *intersection* between the line and the plane.

The line in which two planes cut each other is called the *line of intersection*.

It is often necessary to determine the true form of the straight or curved lines of intersection in representing drawings correctly or in developing sheet metal work.

DEVELOPMENTS. If the surface of an object be unwrapped upon a plane until every part of that surface is in contact with the plane, in its true size, the surface obtained is called a *development* of the surface of the object.

69. Given the hexagon prism shown in Fig. 130, cut by two planes. Draw the development of the lower half of the prism shown in the elevation as follows:

1st. Draw the cutting plane at an angle of 30° so that the prism is divided into two equal parts.

2d. Draw a turned section on an auxiliary plane shown at 1, 2, 3, 4.

3d. Lay out the development of the lower half of the prism thus:

To the right of the elevation in Fig. 130, prolong the base

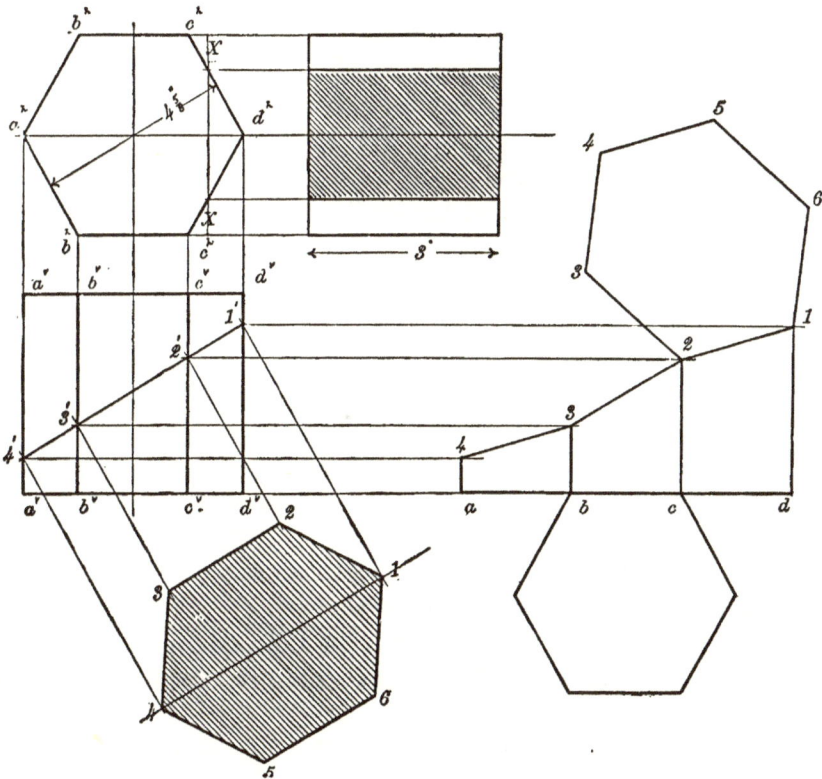

FIG. 130.   FIG. 131.

line indefinitely and lay off upon it the distances *ab*, *bc*, *cd*, etc., Fig. 131, each equal in length to a side of the hexagon. At these points erect perpendiculars, and through the points 1 2′3′4′ draw horizontal lines intersecting the perpendiculars in 4, 3, 2, 1, etc. At *bc* draw the hexagon $a^h b^h b^h$, $c^h c^h$, d^h of the last problem for the base, and at 1, 2 draw the section 1, 2, 3, 4, 5, 6 for the top.

Fig. 131 shows three sides of the lower half of the prism. The remaining three may be drawn in a similar manner.

PLATE 15. This plate will consist of sections and developments. The problems are to be worked out in fine pencil lines, using the 6H pencil. Each problem when drawn should be submitted for criticism. The student should not begin a new problem until the preceding has been approved and checked off. When the drawings of all the problems have been approved, the plate should be cleaned with the art gum and all lines strengthened with the 4H pencil. The title of this plate will be " Sections and Developments." Ink only problem numbers, title and border lines.

PROB. 66. *Draw the elevation and plan of the pentagonal prism shown in Fig. 132.*

Pass a cutting plane B dividing the elevation of the prism into two equal parts at an angle of 30° and develop the lower half of the prism.

1st. Project the true form and size of the section cut by Plane *B* upon an auxiliary plane and draw the development of the lower half in a similar manner to that shown in Art. 69, Fig. 131, for a hexagonal prism.

2d. Add the plane of the base and the plane of the section at *B* in their true sizes to the development.

3d. Draw the trace of Plane *A* in the position shown in the plan, Fig. 132, conceive the part to the right removed and draw the section cut by Plane *A*.

PROB. 67. Given elevation and plan of pyramid shown in Fig. 133. Pass cutting planes, one to cut the elevation at an angle of 45° and another perpendicular to *H* about in the positions shown by the traces.

1st. Draw the center lines for the elevation and plan of the pyramid.

2d. Construct the outline of the plan to dimensions and join the four corners by diagonal lines.

3d. From the point of intersection between the diagonals draw the axis of the pyramid and construct the elevation.

4th. Draw the trace of the cutting plane through the elevation at an angle of 45° and draw the *H* projection of the section.

5th. Draw the true size of the section projected in lines perpendicular to the trace of the plane.

6th. Draw the full development of the lower half of the pyramid including the base and top.

7th. Prolong the cutting plane through the *H* projection to cut the *V* projection and draw the true size of the section to the right of the *H* projection of the pyramid.

FIG. 132.

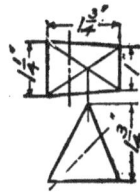

FIG. 133.

70. Fig. 134. To draw the projections of a right cylinder 3″ diameter and 3″ long. (1) When its axis is perpendicular to the H.P. (2) Draw the true form of a section of the cylinder, when cut by a plane perpendicular to the V.P. making an angle of 30° with the H.P. (3) Draw a development of the upper part of the cylinder.

For the plan of the first condition, describe the circle 1′, 2′, etc., with a radius = 1½″ and from it project the elevation, which will be a square of 3″ sides.

For the second condition: Let 1, 7 be the trace of the cutting plane, making the point 7, ½″ from the top of the cylinder. Divide the circle into 12 equal parts and let fall perpendiculars through these divisions to the line of section, cutting it in the points 1, 2, 3, 4, etc. Parallel to the line of section 1, 7 draw 1″ 7″ at a convenient distance from it, and through the points 1, 2, 3, 4, etc., draw perpendiculars to 1, 7, intersecting and extending beyond 1″ 7″. Lay off on these perpendiculars the distances 6″ 8″ = 6′ 8′, and

$5''\ 9''=5'\ 9'$, etc., and through the points $2''$, $3''$, $4''$, etc., describe the ellipse.

For the development: In line with the top of the elevation draw the line $g'g''$ equal in length to the circumference of the circle, and divide it into 12 equal parts a', b', etc., a', b'', etc. Through these points drop perpendiculars and through the points 1, 2, 3, etc., draw horizontals intersecting the perpendiculars in the points 1, 2, 3, etc., and through these points draw a curve.

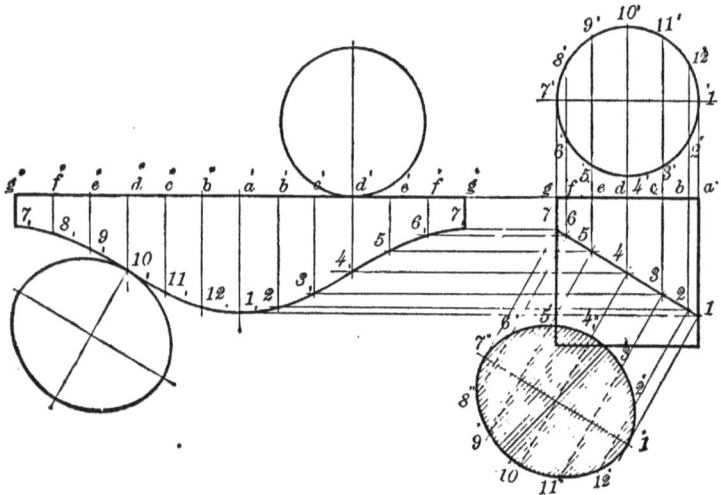

FIG. 134.

Tangent to any point on the straight line draw a $3''$ circle for the top of the cylinder and tangent to any suitable point on the curve transfer a tracing of the ellipse.

PROB. 68. *Draw the projections of a right cylinder $4''$ diameter and $4\frac{1}{4}''$ long. Cut by a plane making an angle of $30°$ with H and dividing the cylinder into two equal parts.*

Draw the true size of the section cut by the plane and develop the lower half of the cylinder in a similar manner to that shown in Fig. 134.

71. FIG. 135. To draw the development of the half of a truncated cone, given the plan and elevation of the cone.

Divide the semicircle of the plan into any number of parts, then with A as center and A1 as radius, draw an arc and lay off upon it from the point 1 the divisions of the semicircle from 1 to 9, draw 9A. Then with center A and radius AB draw the arc BC. 1BC9 is the development of the half of the cone approximately.

PROB. 69. *Given a. right `circular cone whose base is* 3″

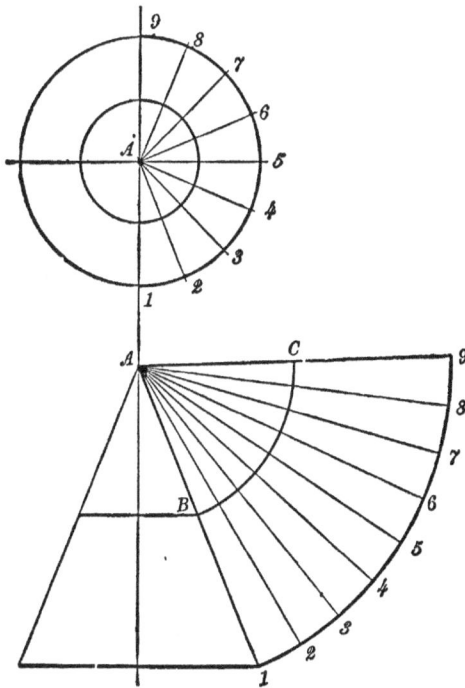

FIG. 135.

diameter and height 4″ *intersected by a V projecting plane. The trace of the plane to pass through the center of the axis and make* 30° *with H.*

Find the true size of the section and lay out the full development of the lower part of the cone. Scale: 6″ = 1′.

72. FIG. 136. To draw the projections of **a right cone** 7″ high, with a base 6″ in diameter, pierced by a **right cylinder** 2″ in diameter and 5″ long their axes intersecting at

right angles 3″ above the base of the cone and parallel to
V.P. Draw first the plan of the cone with a radius = 3″.

At a convenient distance below the plan draw the elevation to the dimensions required.

Three inches above the base of the cone draw the center
line of the cylinder CD, and about it construct the elevation
of the cylinder, which will appear as a rectangle 2″ wide and
$2\frac{1}{2}$″ each side of the axis of the cone. The half only appears
in the figure.

To project the curves of intersection between the cylinder
and cone in the plan and elevation: Draw to the right of
the cylinder on the same center line a semicircle with a radius
equal that of the cylinder. Divide the semicircle into any
number of parts, as 1, 2, 3, 4, etc. Through 1, 1 draw the
perpendicular A'' 1″ equal in length to the height of the cone,
and through A'' draw the line A'' 4″ tangent to the semicircle at the point 4, and through the other divisions of the
semicircle draw lines from A'' to the line 1″ 4″, meeting it
in the points 3″ 2″.

From all points on the line 1″ 4″, viz., 1″, 2″, 3″, 4″,
erect perpendiculars to the center line of the plan, cutting
it in the points 1_1″, 2_1″, 3_1″, 4_1″, and with 1_1″ as the center
draw the arcs 2_1″–2, 3_1″–3, 4_1″–4 above the center line of
the plan, and through the points 2, 3, 4, draw horizontals
to intersect the circle of the plan in the points 2′, 3′, 4′,
and lay off the same distances on the other sides of the center
line of the plan in same order, viz., 2′, 3′, 4′. Through each
of these points on the circumference of the circle of the plan
draw radii to its center A', and through the same points also
in the plan let fall perpendiculars to the base of the elevation
of the cone, cutting it in the points 2′, 3′, 4′; and from the
apex A of the elevation of the cone draw lines to the points
2′, 3′, 4′ on the base. Horizontal lines drawn through the
points of division 2, 3, 4, on the semicircle will intersect the
elements A–2′, A–3′, A–4′ of the cone in the points 2′, 3′, 4′;
these will be points in the elevation of the curve of intersection between the cylinder and the cone.

The plan of the curve is found by erecting perpendiculars
through the points in the elevation of the curve to intersect
the radial lines of the plan in correspondingly figured points,
through which trace the curve, as shown. Repeat for the
other half of the curve.

FIG. 137. To draw the development of the half cone,
showing the hole penetrated by the cylinder.

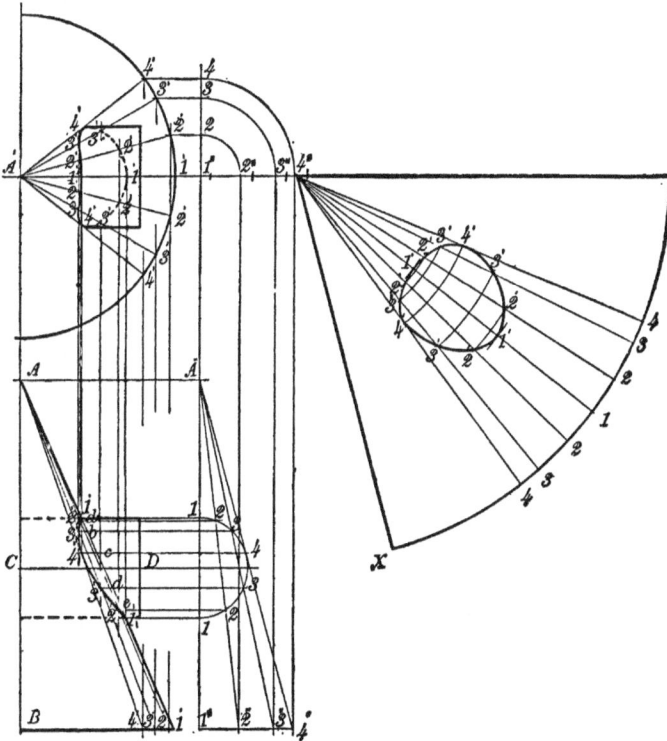

FIG. 136.   FIG. 137.

With center $4_1''$, and element $A1'$ of the cone, Fig. 136,
as radius, describe an arc equal in length to the semicircle
of the base of the cone. Bisect it in the line $4_1''$, 1, and on
each side of the point 1 lay off the distances 2, 3, 4, equal
to the divisions of the arc in the plan Fig. 136, and from these
points draw lines to $4''$, the center of the arc. Then with
radii $A-a$, b, 1, d, e, respectively, on the elevation Fig. 136,

and center $4_1''$ draw arcs intersecting the lines drawn from the arc XX to its center $4_1''$. Through the points of intersection draw the curve as shown by Fig. 137.

PLATE 16. This plate will consist of lines of intersection and developments. Arrange drawings on plate to best advantage.

PROB. 70. *Draw the projections of a right circular cone with a 3″ base and 4″ high, pierced by a right cylinder 1½″ diameter. The center line of the cylinder to be 1¼″ above the base of the cone.*

1st. Find the curve of intersection between the cylinder and the cone according to Art. 72.

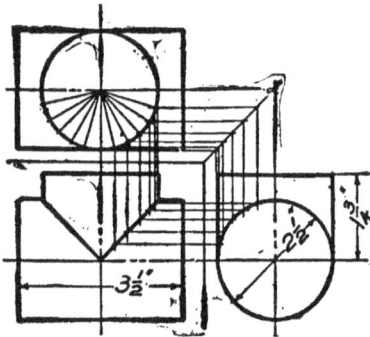 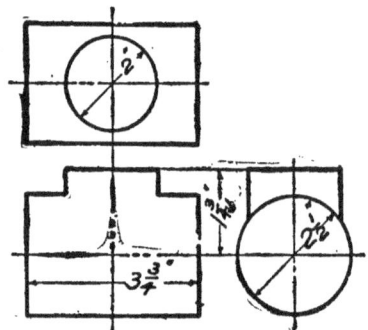

FIG. 138.   FIG. 139.

2d. Draw the development of half of the cone showing the hole made by the penetration of the cylinder. Scale: $12'' = 1'$.

73. **To find the curve of intersection between two right cylinders intersecting each other at right angles.**

Fig. 138 shows three views of two right cylinders of equal diameter, intersecting at right angles.

1st. Divide the semi-circle in the plan into 12 equal parts and from the points on one-quarter of the circumference draw horizontals to meet the bisector of the 90° angle of revolution and from those points drop perpendiculars to cut circumference of the circle in the end view of the other cylinder.

2d. From the same points in the plan drop perpendiculars to meet horizontals drawn from the corresponding points in the end view.

3d. Draw the curve of intersection through the points of intersection between the perpendiculars and horizontals in the front view. The curve in this case will be a straight line because the intersecting cylinders are equal in diameter.

PROB. 71. *Draw the curve of intersection between the two right cylinders shown in Fig. 139, to the dimensions given, and development of small cylinder.*

This line of intersection will be an irregular curve. When the points in the curve are obtained, sketch freehand a light line through them, making a smooth curve, and when finishing the drawing strengthen the line with the 4H pencil and the French curve.

Study Arts. 70 and 73.

PROB. 72. *Given the elevation and plan of the square prism penetrated by a hexagonal prism, Fig. 139a. Draw the development of either the square prism or the hexagonal prism showing the true size of the part remaining of one after the other has been removed.* Scale: to suit.

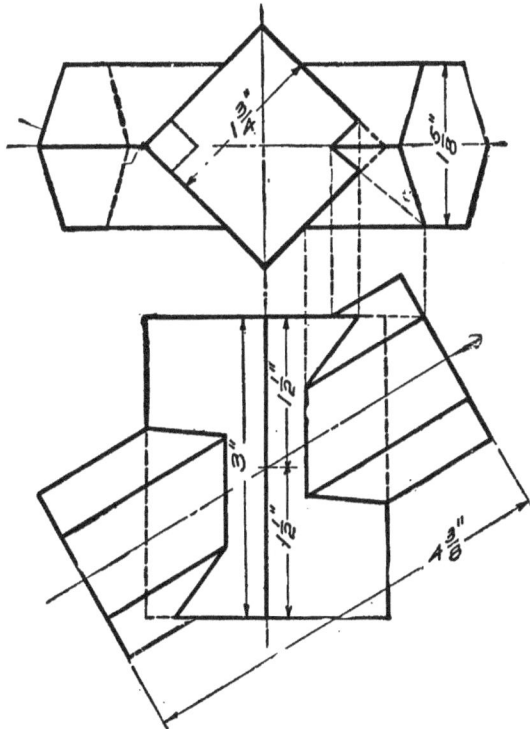

FIG. 139a.

See Appendix to Mechanical Drawing for additional problems in intersections and developments.

CHAPTER VIII

ISOMETRICAL PROJECTION

74. In orthographic projection it is necessary to a correct understanding of an object to have at least two views, a front and end elvation or an elevation and plan, and sometimes even three views are required.

Isometric projection on the other hand shows an object completely with only one view. It is a very convenient system for the workshop. Davidson in his *Projection* calls it the "Perspective of the Workshop." It is more useful than perspective for a working drawing, because, as its name implies ("equal measures") it can be made to any scale and measured like an orthographic drawing. It is, however, mainly employed to represent small objects, or large objects drawn to a small scale, whose main lines are at right angles to each other.

The principles of isometrical projection are founded on a cube resting on its lower front corner, 1, Fig. 140, and its base elevated so that its diagonal AB is parallel to the horizontal plane. Then, if the cube is rotated on the corner 1 until the diagonal AB is at right angles to the vertical plane, i.e., through an angle of 90°, the front elevation will appear as shown at 1, 2, 3, 4, Fig. 140, a regular hexagon.

Now we know that in a regular hexagon, as shown by Fig. 140, the lines $1A$, $A3$, etc., are all equal, and are easily drawn with the 30°×60° triangle. But, although these lines and faces appear to be equal, yet, being inclined to the plane of projection, they are shorter than they would actually be on the cube itself. However, since they all bear the same proportion to the original sizes, they can all be measured with the same scale.

75. To make an isometrical scale.

Draw the half of a square with sides $= 2\frac{1}{2}''$, Fig. 141. These two sides will make the angle of 45° with the horizontal

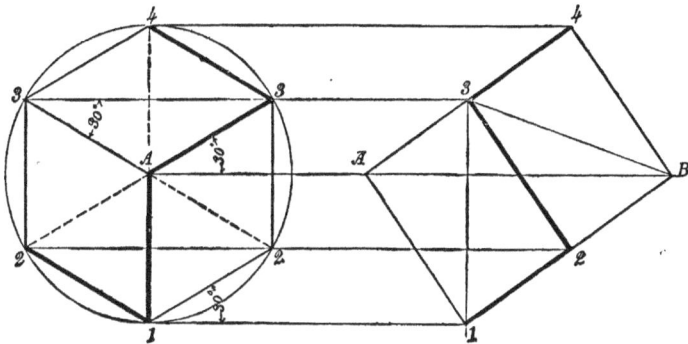

FIG. 140.

Now the sides of the corresponding isometrical squares, we have seen, makes the angle of 30° with the horizontal, so we will draw 1, 4, 3, 4, making angles of 30° with 1, 3. The

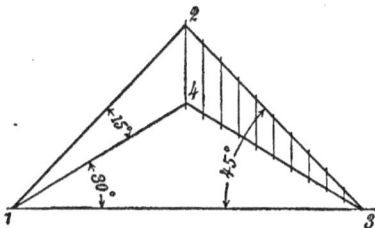

FIG. 141.

difference then between the angle 2, 1, 3 and the angle 4, 1, 3 is 15°, and the proportion of the isometrical projection to the actual object is as the length of the line 3, 2 to the line 3, 4. And if the line 3, 2 be divided into any number of equal parts, and lines be drawn through these divisions parallel to

2, 4 to cut the line 3, 4 in corresponding divisions, these
will divide 3, 4 proportionately to 3, 2.

Now if the divisions on 3, 2 be taken to represent feet
and those on 3, 4 to represent 2 feet, then 3, 4 would be an
isometrical scale of $\frac{1}{2}$.

76. Isometrical Drawing. Since isometrical drawings may
be made to any scale, we may make the isometrical lines of

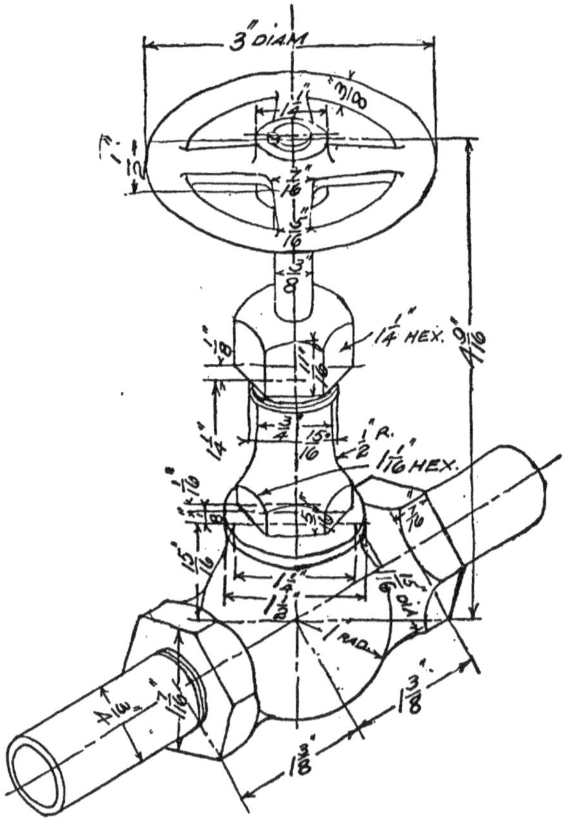

FIG. 142.

the object = their true size. This is a common practice and
precludes the need of a special isometrical scale.

When an object is drawn to its true size the result is
called an *Isometrical Drawing.*

Fig. 142 gives a detail isometric drawing of the $\frac{3}{4}''$ globe valve shown in the piping layout.

Fig. 143 shows an isometric drawing of an engine connecting rod partly in section.

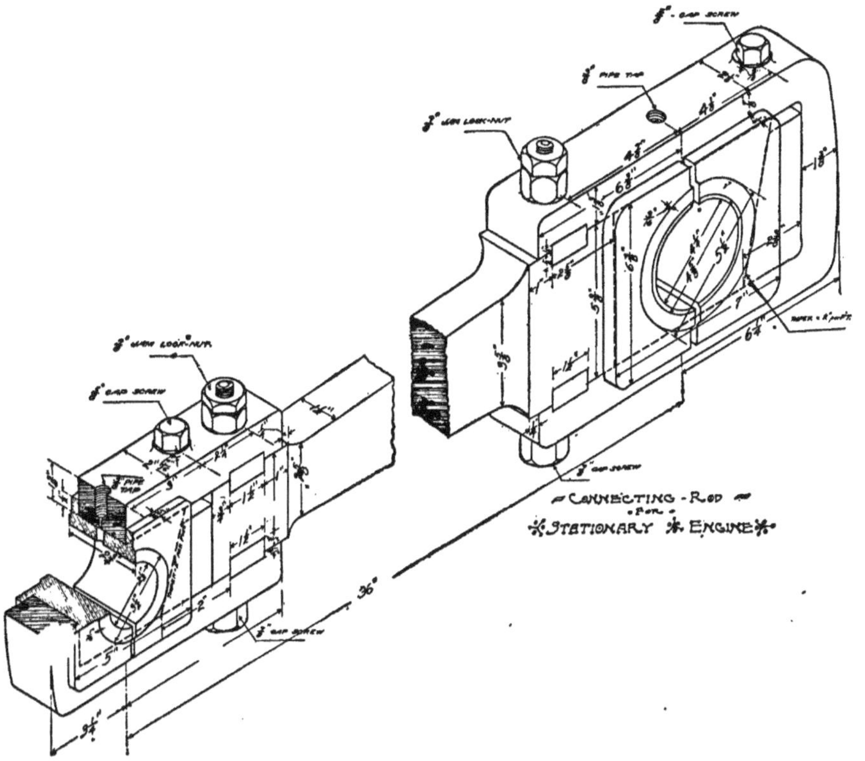

FIG. 143.

Fig. 144 shows an isometrical drawing made by the Murray Engine Works. It represents in one view the layout of a power plant.

Fig. 145 shows a detail isometric drawing of the layout of the steam piping for the above power plant.

Fig. 146 shows an isometric drawing of an engine connecting rod.

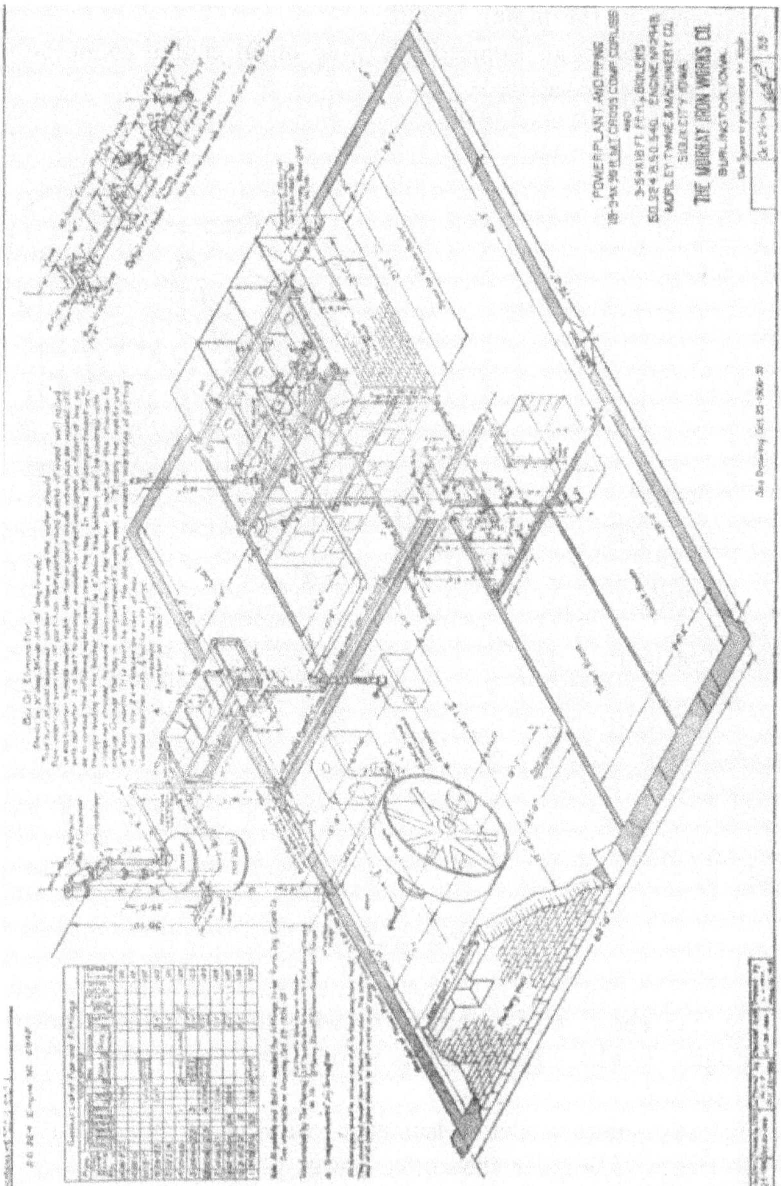

FIG. 144.

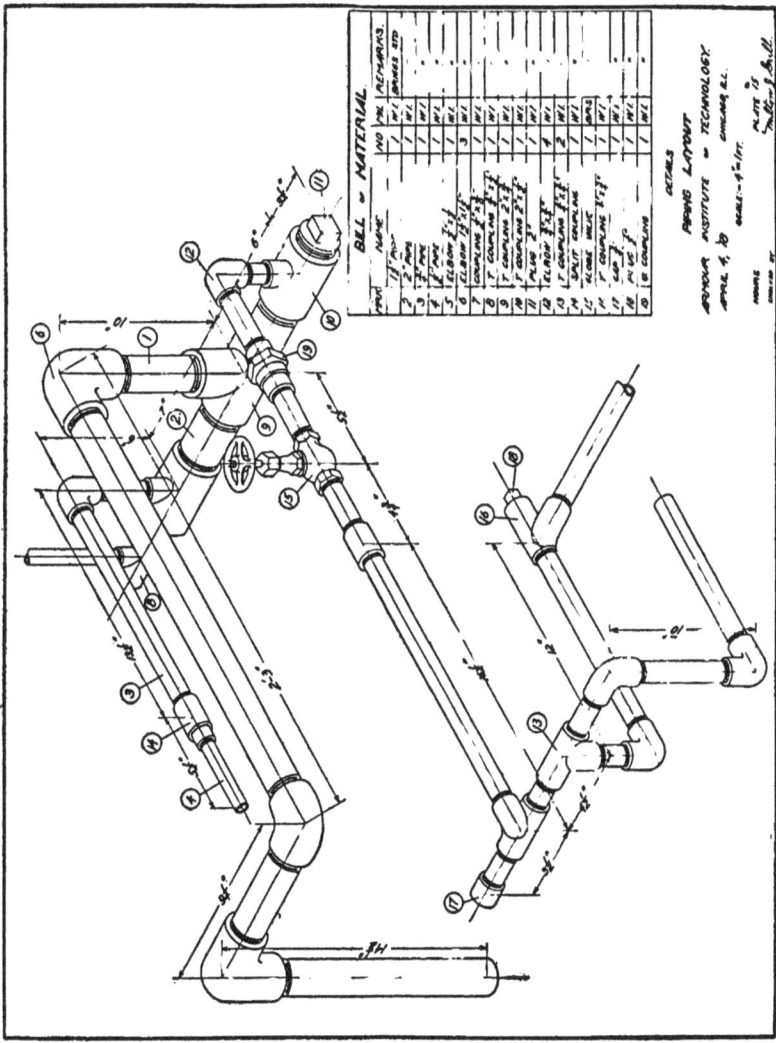

FIG. 145.

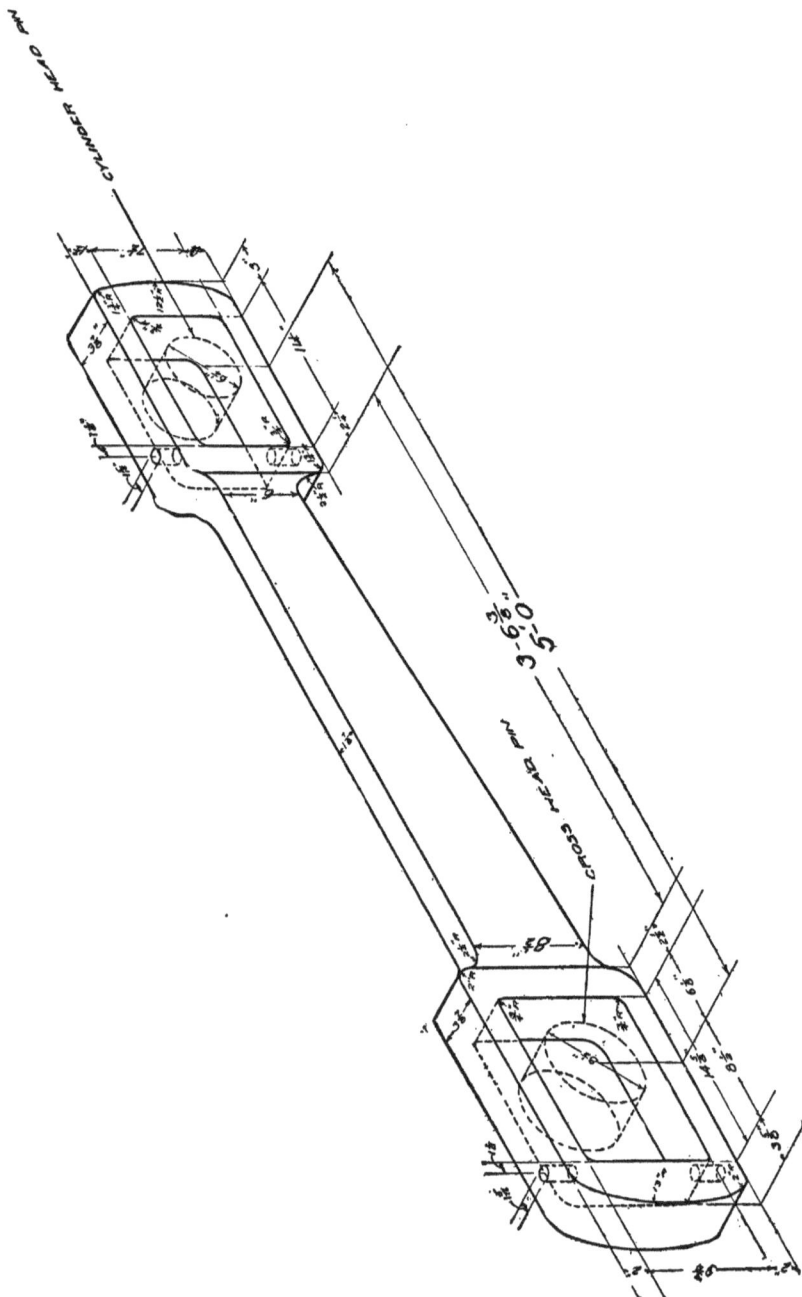

FIG. 146.

77. To make the isometrical drawing of a two-armed cross standing on a square pedestal.

Begin by drawing a center line AB, Fig. 147, and from the point A draw AC and AD, making an angle of 30° with the

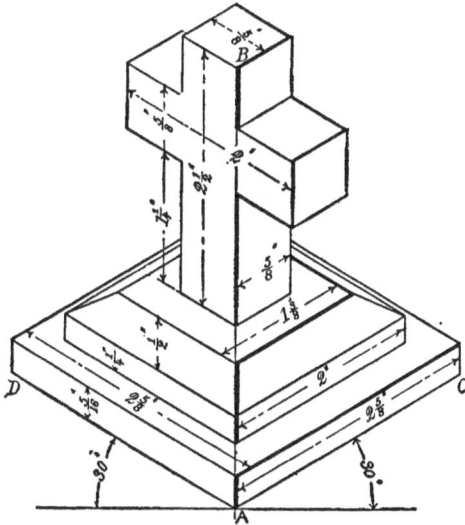

FIG. 147.

horizontal. Measure from A on the center line AB a distance $= \frac{5}{16}''$, and draw lines parallel to AC, AD; make AC and AD $2\frac{5}{8}''$ long and erect a perpendicular at D and C, completing the two front sides of the base, etc.

Note: Measurements should always be made along an isometric line.

78. To Draw the Isometric Projection of a Circle on a Horizontal Plane.

Draw a $2''$ circle and describe an orthographic square about it, using the 45° triangle, Fig. 148.

Draw the diagonals 1, 2, 3, 4 and the diameters 5, 6, 7, 8 at right angles to each other.

Now from the points 1 and 2 draw lines 1A, 1B and 2A, 2B, making angles of 30° with the horizontal diagonal 1, 2.

And through the center O draw CD and EF at right angles to the isometrical square.

The points CD, EF, and GH will be points in the curve of the projected isometrical circle, which will be an ellipse.

The ellipse may be drawn sufficiently accurate as follows:

With center B and radius BC describe the arc CF and extend it a little beyond the points C and F, and with center A and same radius describe a similar arc, then with a radius which may readily be found by trial, draw arcs through G and H. These smaller arcs are not to be drawn tangent to the sides of the square at the points G, D, E, or F, but

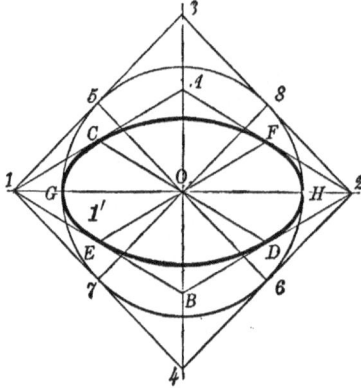

FIG. 148.

are to be drawn tangent to the larger arcs extended beyond the points C, D, E, and F, as stated.

This gives a close approximation to a true ellipse and maintains the true length of the major axis, which is very desirable, because the major axis must always remain the same length whatever the angle of the plane may be in which the circle is projected.

An isometric circle drawn in this way presents a neat and satisfying appearance.

79. To Find the Small Radius V of the isometric circle. With OC, Fig. 148, as radius and point 1 as center, cut the major axis of the ellipse in $1'$. $1'$ is the center of radius v of the approximate ellipse or isometric circle.

The use of this method for determining the radius precludes the need of drawing the orthographic circle or square when making an isometric " drawing " of a circle.

80. To Make the Isometric Drawing of a Hexagonal Bolt Head on a Vertical Plane. Fig. 149.

1st. Draw the center line C of the bolt from left to right, making the angle of 30° with the horizontal and another line

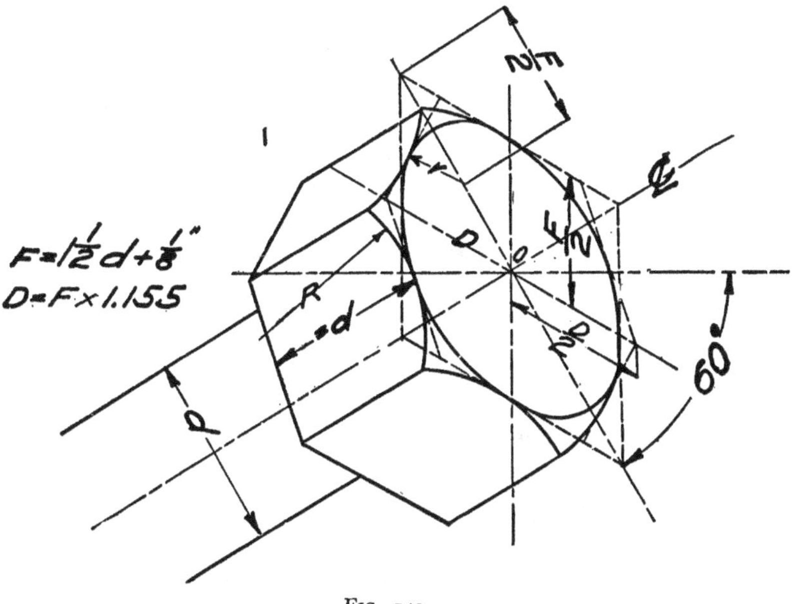

$$F = \tfrac{1}{2}d + \tfrac{1}{8}''$$
$$D = F \times 1.155$$

FIG. 149.

D at 30° from right to left, cutting the first line at the center of the hexagon and through the same center draw a perpendicular, as shown in Fig. 149.

2d. Draw the isometric square with sides equal to $F = 1\tfrac{1}{2}d + \tfrac{1}{8}''$ = the distance across the flats of the hexagonal head.

3d. Draw the isometric inscribed circle in the usual way and describe the hexagon about it as shown in the figure, laying off the long diameter of the hexagon on the center line D, making $D = F \times 1.155$.

4th. Make radius R equal to the width of the face and all the edges of the hexagonal head equal to each other in

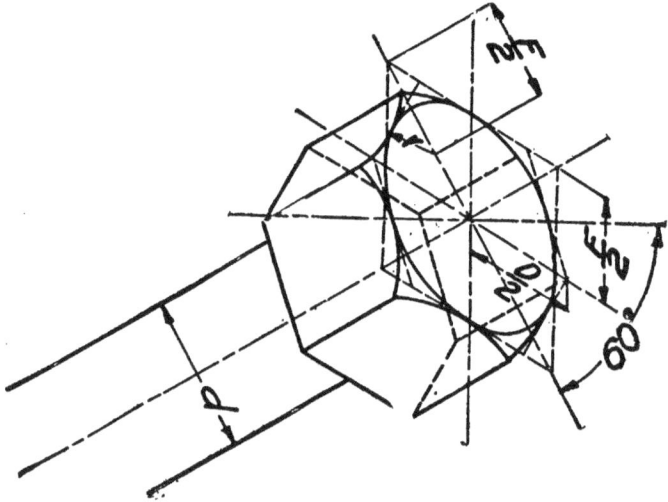

FIG. 150.

length. The chamfer curves on the other faces may be easily drawn by trial radii.

Fig. 150 shows the isometric "*projection*" of the same bolt and bolthead.

81. To Lay Off an Angle on an Isometric Circle.

1st. Construct the ellipse from the orthographic circle as explained in Art. 78.

2d. Lay off the true angle A as shown in Fig. 151, by a dotted line and when this line cuts the orthographic circle erect a perpendicular to the ellipse or isometric circle and through that point draw a line from the center making, the corresponding isometric angle a.

This method can be used in determining the points of the hexagon as shown at Fig. 152 as follows:

Given the orthographic and isometric circles, draw the broken line through the center of the circle, making an angle of 60° with the horizontal (since there are six angles of 60°

each in a hexagon), Fig. 152, and produce it to cut the orthographic circle, as shown in a point. Revolve that point back

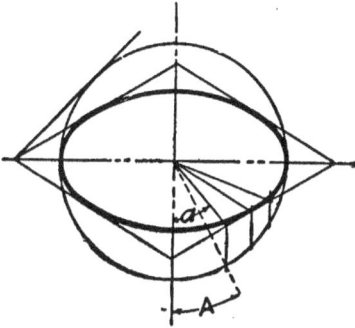

FIG. 151.

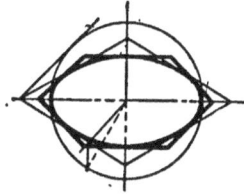

FIG. 152.

into the isometric circle by drawing a line through the point parallel to the axis to intersect the isometric circle in a corresponding point. From the center draw a solid line through the latter point and produce it to intersect a horizontal tangent to the isometric circle. The point of intersection is a point of one of the angles of the isometric hexagon. The other angles may be found in a similar manner.

82. **To lay off an angle from a corner of the isometrical cube.**

Construct an orthographic square of any convenient size as shown in Fig. 153 and draw the required angle *AOB*.

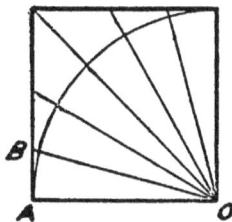

FIG. 153.

From the corner of the isometrical cube where the angle is to be drawn lay off along the side a distance equal to *OA* of the orthographic square and erect a perpendicular at A. Step off the distance AB and draw OB the angle required. Any other angle may be drawn in similar manner.

PLATE 17

PROB. 73. *Make the isometrical drawing of a $2\frac{1}{4}''$ cube. Draw a $2\frac{1}{4}''$ isometric circle on the upper face by the method shown in Fig. 148, page 136. From the lower left-hand corner of the right-hand face lay off angles of $15°$, $30°$, and $45°$. Use method shown in Fig. 153.*

PROB. 74. *Make the isometrical drawing of the hollow cube with a hollow block on each face as shown in Fig. 154.*

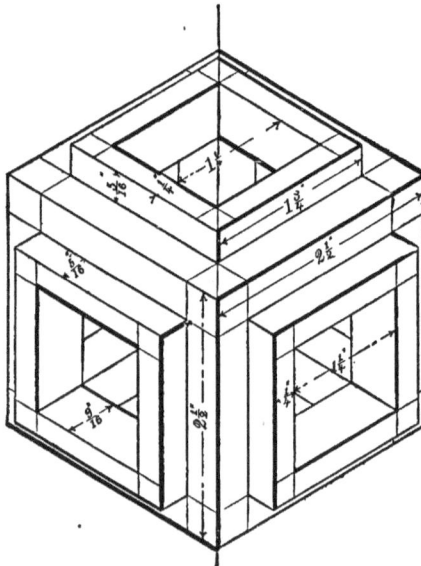

FIG. 154.

PROB. 75. *Make the isometrical drawing of a hexagonal headed bolt and nut on a horizontal plane as shown in Fig. 155.*

Follow directions given in Art. 80 and make drawing similar to Fig. 149.   Bolt to be $1\frac{1}{4}'' \times 4\frac{1}{2}''$.

PROB. 76, Fig. 156. *Make the isometrical drawing of a pentagonal prism. Sides* $1\frac{1}{2}''$. *Height of prism* $2\frac{1}{2}''$.

1st. Construct pentagon of $1\frac{1}{2}''$ sides by Prob. 50, page 56.

2d. Through the point A on the vertical axis of the pentagon

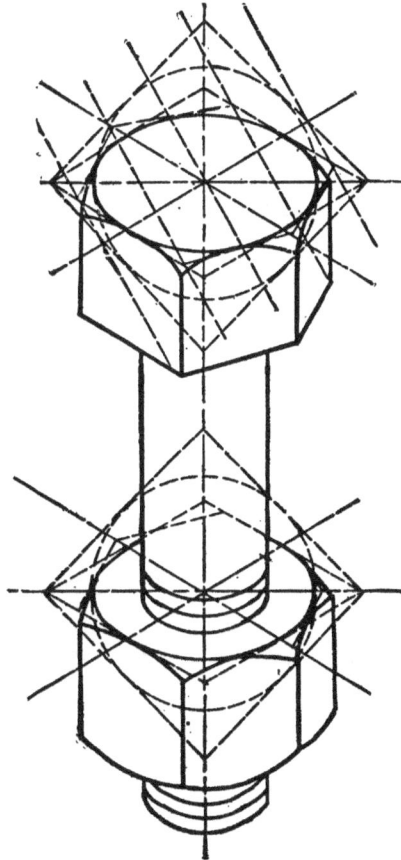

FIG. 155.

draw the isometric line $1'-5'$ at the angle of 30° to the left and A, $3'$, to the right.

3d. With center A and A, 1 as radius cut points $1'$ and $5'$.

4th. Draw through o the isometric line o_1o' and through c the line $2'$, $4'$.

5th. Through the point 3 draw 3, $3'$ to intersect A, $3'$ in the

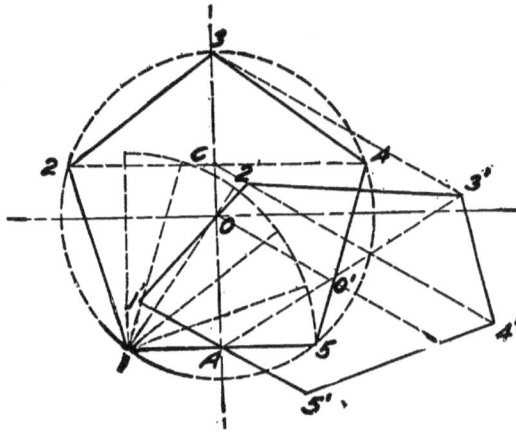

FIG. 156.

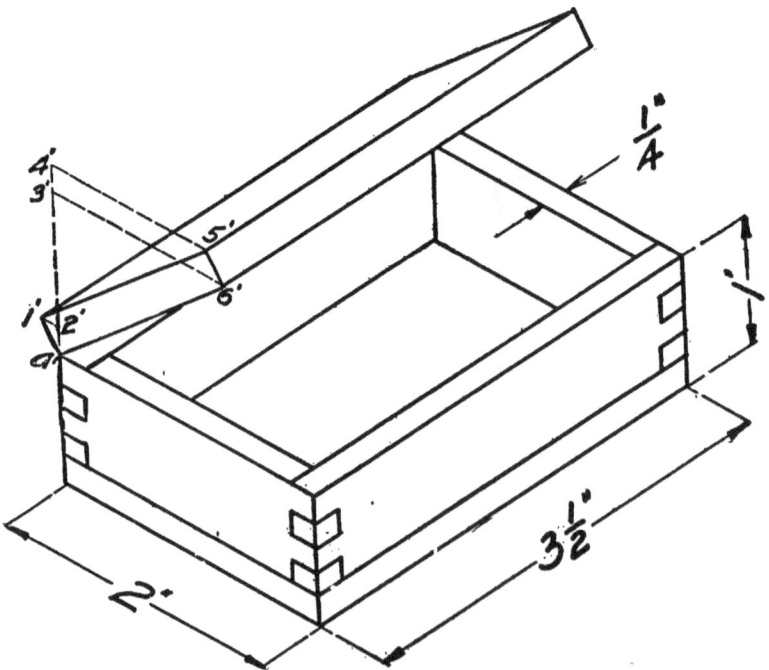

FIG. 157.

point 3'. Join the points 1', 2', 3', 4' and 5', with straight lines completing the isometric drawing of the pentagon. 1', 5' is the only isometric line in the pentagon and measures the exact size of the side, viz., $1\frac{1}{2}''$.

6th. Finish the construction of the prism by drawing the edges of the prism downward $2\frac{1}{2}''$ long and connect with straight lines.

PROB. 77.   *Make the isometrical drawing of the tool box, shown at Fig. 157, to the dimensions given.*

1st. Draw the box in isometric as shown.

2d. To draw the isometric of the cover when open at the angle of 30°, draw the end view of the box orthographically as shown setting the cover at the actual angle of 30°. Fig. 158.

3d. Draw the broken lines 1, 2; 3, 6 and 4, 5 horizontally through the corners of the cover, as shown.

4th. At the point a, Fig. 158, erect the perpendicular broken line a, 4 and take the distance a, 2, Fig. 158, and lay off a', 2',

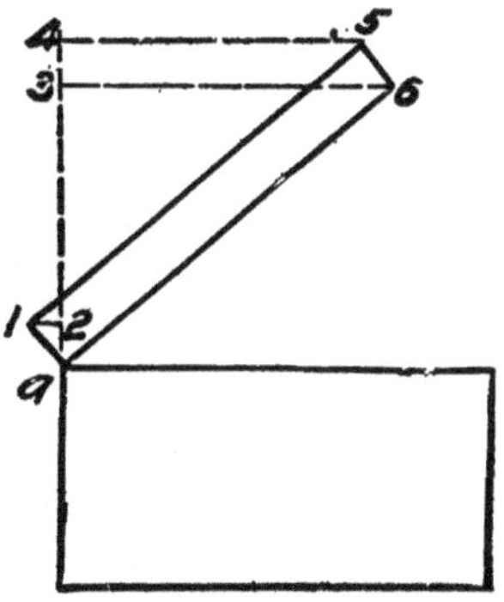

FIG. 158.

Fig. 157.   Draw 1', 2', Fig. 157, at the angle of 30°. With the distance 1, 2, Fig. 158, lay off 1', 2', Fig. 157.

5th. Take the distance a, 3, Fig. 158, and lay off a', 3', Fig. 157, and make 3', 6', equal to 3, 6.

6th. Make a', 4' equal to a, 4 and 4', 5' equal to 4, 5 and join the points a', 1', 5', and 6' with straight lines.

7th. Complete the isometric drawing of the cover and place dimensions as shown.

PROB. 78. *Find the line of intersection on the 2" by $1\frac{1}{2}$" T coupling. Fig. 159.*

1st. Draw the coupling as shown.

2d. On the center line of o, o of the small cylinder describe a semicircle and divide it into 12 equal parts, from o to 6 and from 6 to o.

3d. On the right-hand end of the large cylinder, at the center C erect a perpendicular C, O'. O' is the top of the cylinder.

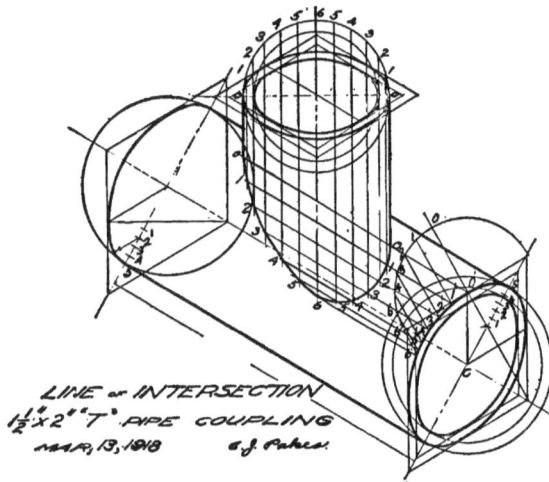

FIG. 159.

4th. At point O' draw a perpendicular to the isometric square and describe the same semicircle as on the small cylinder and divide it from o to 6 as before. Through these points draw lines parallel to the perpendicular $o_1 o'$, cutting the large ellipse in points o_1' $1_1'$ $2'$, etc.

5th. Through points o_1' 1_1 '2', etc., draw lines parallel to the axis of the large cylinder to intersect the corresponding perpendicular lines from the same points on the small cylinder.

6th. Through the resulting points draw the required curve of intersection.

CHAPTER IX

WORKING DRAWINGS

83. A working drawing is made to convey all the necessary information from the drafting room to the shop to enable the workmen to correctly make a pattern for castings, to make parts to be forged in wrought iron or soft steel, to make parts to be made in sheet metal. To give the necessary information to the machine shop, to do the finishing and to the fitting shop to properly assemble the different parts into the whole machine.

84. The making of a working drawing may be divided into consecutive steps as follows:

1st. *The selection of the proper size of detail paper for the drawing or drawings to be made.*

In most drafting rooms standard sizes are used for the different details to be drawn; such as $9'' \times 12''$, $12'' \times 18''$, $18'' \times 24''$, etc.

For the purposes of this course all our drawing are made on sheets $15'' \times 20''$ over all.

2d. *Drawing the standard border line.*

The border line will be the same for all drawings $1\frac{1}{2}''$ at left-hand end and $\frac{1}{2}''$ at top and bottom and right-hand end.

The width of the border line should be $\frac{1}{16}''$.

3d. *The arrangement of the different details to be placed on the sheet including the necessary views of each part.*

For each plate a definite number of problems will be assigned.

The draftsman should make some calculations to determine approximately the space required for essential views of the different objects together with suitable spaces between views and between the different problems, having in mind the space required itle and bill of material.

Different views of an object should be spaced closer together than the space allowed between different objects.

4. *The style, location, drawing and application of title, bill of material, notes and dimension figures.*

1. The style of the title is given in Fig. 160. The guide lines should be carefully drawn in fine light lines with the 6H pencil according to the spacing given.

2d. The bottom line of the title should be about $\frac{1}{16}$" inside of the border line in the lower right-hand corner of the plate.

3d. The longest line of the title should be lettered first and a light vertical line drawn through the center of it so that the remainder of the title may be balanced with reference to that center line.

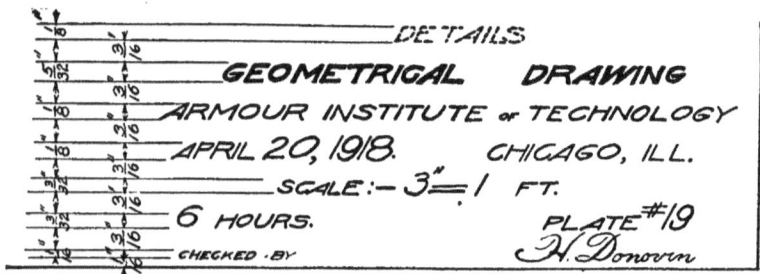

FIG. 160.

4. The *name* line of the standard title should be inked with heavy lines to give prominence.

Ink first in ordinary lines with ball-point pen No. 516 and then strengthen the lines of all the letters of the words in the main title, using the Gillott fine pen No. 303.

In drawing the bill of material table, observe the dimensions given in Fig. 161.

Make the name space to suit the longest name to be recorded, using appropriate abbreviation, consulting Section 30.

The bottom line of the bill of material should be placed $\frac{1}{4}$" above the top line of the title.

In the column for " Number " of pieces any number may generally be recorded.

In the column for " Remarks " state any suitable information for the shop not given on the drawing such a patt., style, etc.

5. *Procedure in Making Pencil Drawing.*

1st. When the border lines, space for title and bill of material have been laid off, then on the remaining space on the plate lay out the center lines of the different views of the drawings to be made.

2d. Begin at the upper left-hand corner of the plate and draw the different views of the first object to be drawn in fine pencil lines. Make the details complete and when all the views have been carefully drawn in fine narrow lines with the 6H pencil properly sharpened the drawing should be submitted for approval. If it is all right or when it is made right the instructor will check it off and the next problem may be drawn in a similar

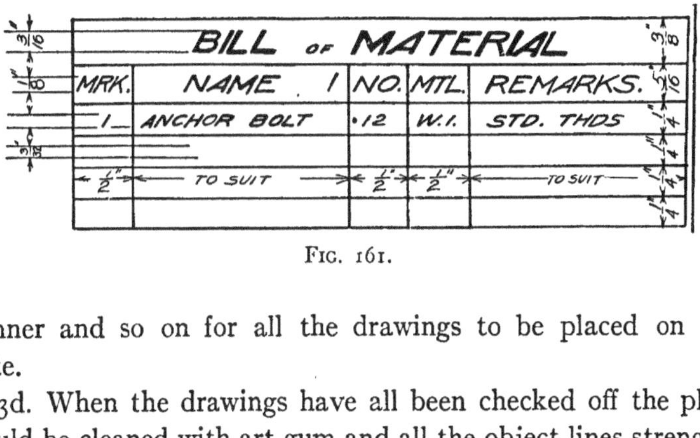

FIG. 161.

manner and so on for all the drawings to be placed on the plate.

3d. When the drawings have all been checked off the plate should be cleaned with art gum and all the object lines strengthened with the 4H pencil properly sharpened to give a strong, clear line.

4th. The dimension lines are next to be applied, very narrow, with the 6H pencil, and also guide lines for information notes.

5th. Put on all dimensions with the 4H pencil, beginning with the left-hand arrow point, then the dimension and sign of inches and then the right-hand arrow point. This procedure in placing dimensions should be observed at all times so as to avoid leaving off arrow points, sign of inches, figures, etc.

6th. Letter all notes, title, bill of material and last of all draw the crosshatching lines on section surfaces. The distance between cross-hatch lines should be $\frac{1}{16}''$.

7th. The student should next check his drawing for errors that it may be as correct as possible before submitting for final approval.

6. *Tracing and Blue Printing.*

1. Drawings these days are often made directly on the dull side of the tracing cloth or on tracing paper and then inked. There is not much time saved in this method except when prints are to be gotten out in a hurry without much regard for their appearance, then the prints are made directly from the pencil drawing without inking. In this way prints can be made quickly, but unless the exposure is accurately made the result is a poor print with a liability to errors.

2. In making a blue print from a tracing, the *back* of the tracing is placed against the sensitized surface of the prepared print paper and exposed in a printing machine with the face of the tracing toward the light.

3. The time of exposure must be learned by trial, because there is often much difference in the sensitiveness of the print paper and also in the transparency or opaqueness of the tracing medium.

4. When the exposure to the light is completed the print may be developed by immersing in a tank of clear water, face up. Lave the water over the face of the print a few times with the hand and then hang up to dry.

When dry the print should be trimmed to the exact size of the drawing plate, in this case $15''\times20''$.

PLATE 18

This plate is to consist of one or more of the following problems and is to be drawn in pencil first according to the directions given in Art. 84, and when approved is to be traced in ink on the dull side of tracing cloth.

A tracing in ink should be begun by inking first the small arcs of circles with the spring bow pen, then the larger arcs and

circles with the compass pen and all irregular curves with the
French curve, after which the straight lines should be inked
with the straight line pen and the figures and lettering with the
writing pen, ball-point No. 516.

PROB. 79, FIG. 162. *Draw a* **front** *view and* **plan** *of the connecting rod shown in Fig. 162. Scale: 3″=1′. See Fig. 146.*

1st. Begin by drawing the *H* center line of the plan 1⅛″
below the top border line and the *H* center line of the elevation
3½″ below the center line of the plan.

2d. Draw the *V* center line through the center of the crank
pin circle shown in adjacent part lines, and the *V* center line

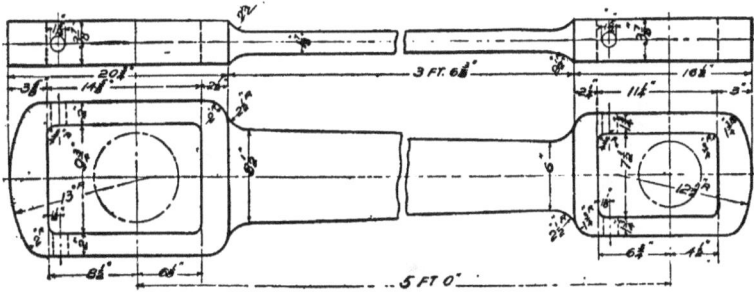

FIG. 162.

through the center of the crosshead pin circle also shown in
adjacent part line at the right-hand end of the connecting rod.

4th. Lay off the widths and lengths of the left stub end and
then of the right-hand stub end according to the dimensions given.

5th. To determine the slope of the rod body. The plate is
not long enough to draw the rod to its true length so it is broken
in the middle as shown.

From the 8½″ line at the left-hand end of the rod body lay
off 3 ft. 6⅜″ to the right on center line at that point draw a perpendicular line intersecting the center line of the rod and lay
off 3″ on each side of the center line and draw the taper rod
through the 8½ and 6″ points as long as shown to the break.

6th. Repeat the process to the left for the right-hand taper.
Study carefully Art. 84 for the detail methods to be used in
making the complete pencil drawing tracing and blue print.

PROB. 80.   *Make the V, H and Pr projections of the Engine Link Support given in the isometric drawing, Fig.* 163   *Scale:* 4″ = 1′.

1st. Lay out all the *V, H* and *P* center lines in convenient positions on the plate.

2d. Arrange position of *V* section indicated on Fig. 163.

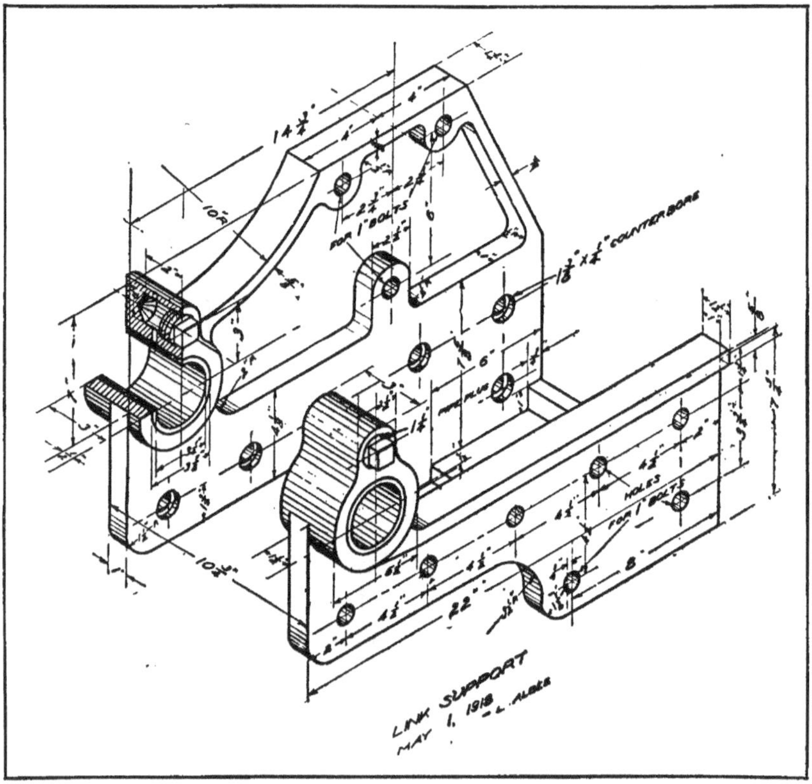

FIG. 163.

3d. Follow instructions given in Art. 84 for the proper completion of the drawing.

Indicate finished surfaces as directed in Art. 29.

PROB. 81.   *Given the elevation and plan of the planer Top Bracket, Fig.* 164. *Draw also the end view, omitting all hidden lines.* *Scale* 6″ = 1′.

Fig. 165 is the isometric of the Top Bracket.

NOTE: The 2″ and 1⅝ circles are to be drawn on the same center line with the upright arm, which is 1⅞″ wide. Fig. 163.

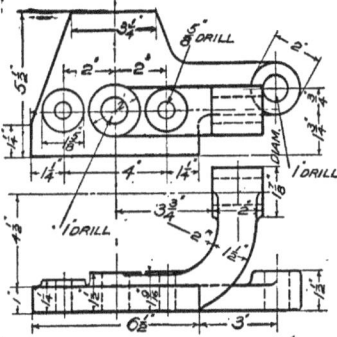

FIG. 164.

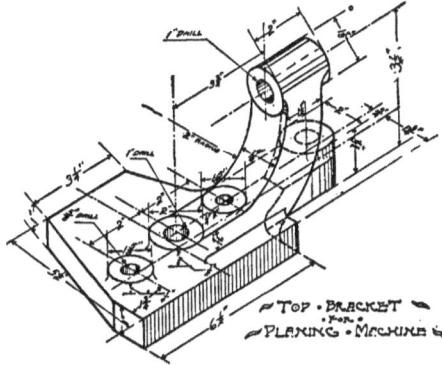

FIG. 165.

PROB. 82. *Draw the Engine Axle shown in Fig. 166. Make cross-sections as shown. Material cold rolled steel (C.R.S.) Scale:* 6″ = 1′.

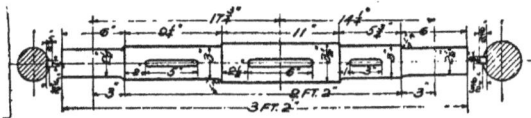

FIG. 166.

PROB. 83. *Given the V and P projection of the automobile axle shown in Fig. 167, draw also a plan of the top. Scale:* 3″ = 1′.

1st. Lay out center lines of the V and P projections shown in Fig. 167, locating them so as to allow for plan above elevation.

2d. Draw center line of plan 1¼″ below upper border line.

3d. Make the drawing of the elevation around the center lines.

4th. Draw the end views similar to those shown, making improvements where possible.

5th. Draw plan from elevation and end view.

6th. When complete in fine pencil lines submit for approval.

7th. When approved in fine lines, clean, strengthen and place dimension lines, dimensions, notes and title according to previous directions.

85. Title Sheet or Set of Drawings. When all the drawings of the course in mechanical drawing have been completed they should be placed together in the order of their making with

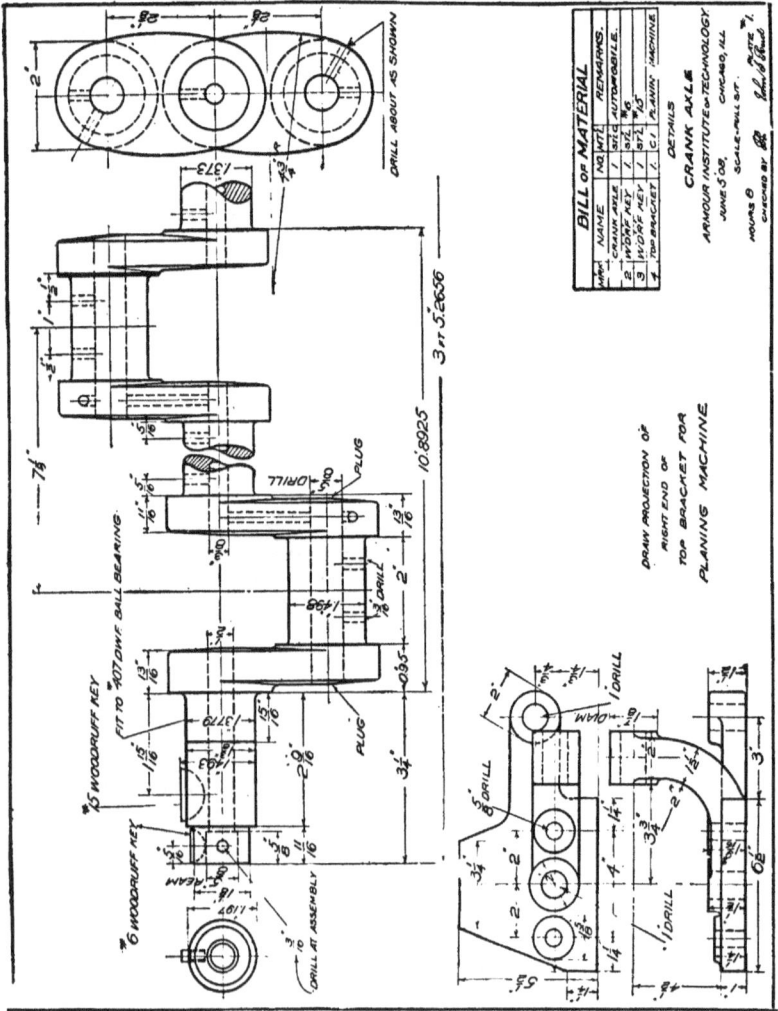

FIG. 167.

first drawing on top and taken to the instructor to be checked up with the records to see that due credit has been given for each drawing.

The drawings should then be bound together with a cover plate design appropriately lettered.

This cover plate is to be placed on top of the drawings and the whole fastened with brass paper binders.

Cover Plate Design

The general style and layout of the lettering is left to the taste of the student except that the main title " Mechanical Drawing " is to be made with the Roman letter, all capitals, shown in Fig. 173 of the Appendix, page 158.

FREEHAND   LETTERING

AND

MECHANICAL DRAWING

PLATES I TO 22 INCLUSIVE

ARMOUR INSTITUTE OF TECHNOLOGY

NOVEMBER 10, 1914   CHICAGO, ILL.

DAVID A PAREIRA

FIG. 168.

Fig. 168 gives a sample title plate showing the heights of letters of the different lines.

Figs. 169 to 172 inclusive are suggestive designs, but the student is urged to make his cover plate with a design of his own as far as possible.

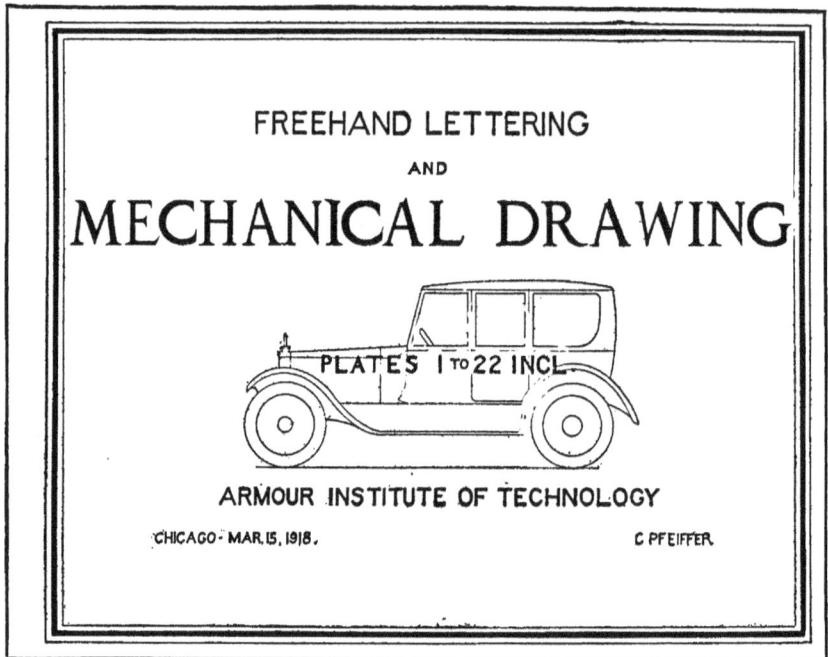

FIG. 169.

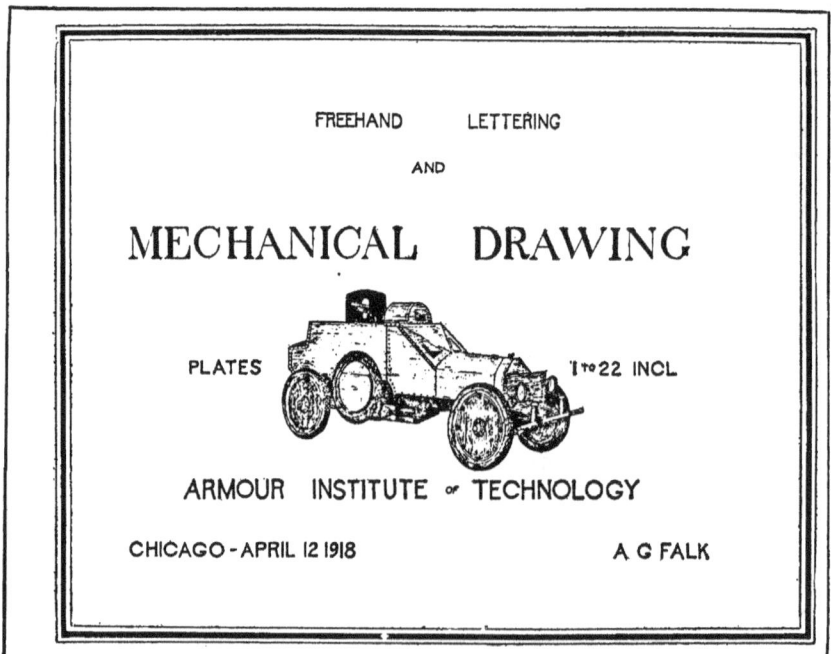

FIG. 170.

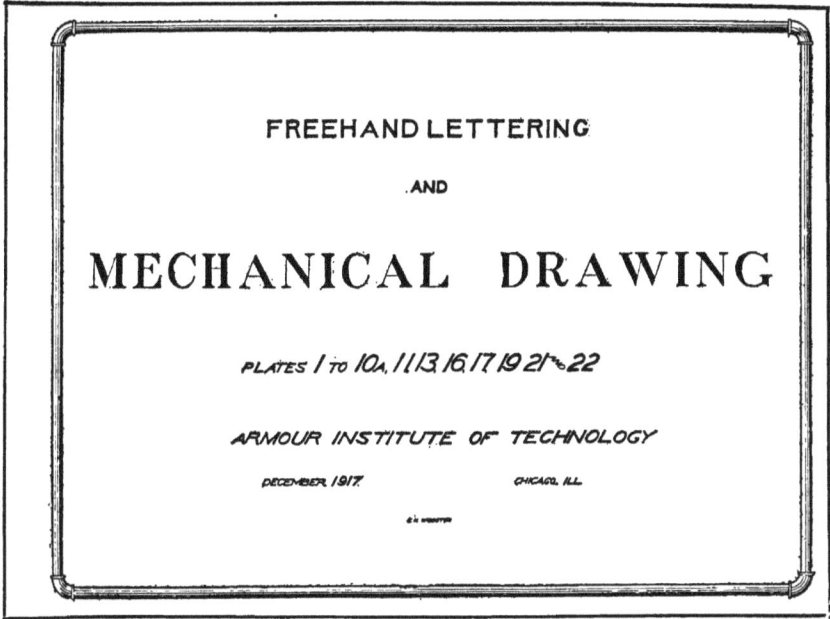

FIG. 171.

FIG. 172.

APPENDIX

REQUIRED COURSE IN MECHANICAL DRAWING

---

86. Lettering Continued. While it is true that the Gothic letter, all capitals, sloping and equal in height is the preferred letter for notes and titles on commercial drawings, there are a few other styles that are sometimes used on special drawings.

FIG. 173 is a carefully proportional Roman letter that is often used for titles of outside cover plates, etc. This letter is difficult to make well freehand, but when made mechanically according to the proportions given in Fig. 173, it presents a fine appearance and can be used in conjunction with other styles. See Art. 85 on cover plates, page 152.

FIG. 174 shows the lower case Roman letter. This letter is not much used in machine drawing.

FIG. 175 shows the upper- and lower-case letters of a vertical Gothic that is sometimes used.

FIGS. 176 and 177 give a style of letter much used by architects both for titles and notes on drawings.

More latitude is allowed to the architectural draftsman in his choice of styles of lettering for notes and titles on working drawings than is given to the machine draftsman. The latter is required to use that style of letter which gives the neatest appearance with a maximum of legibility and requires the least amount of labor and time to construct it; while the former is expected to use a style of letter suggested by the character of the drawing to be named and noted.

FIG. 173.

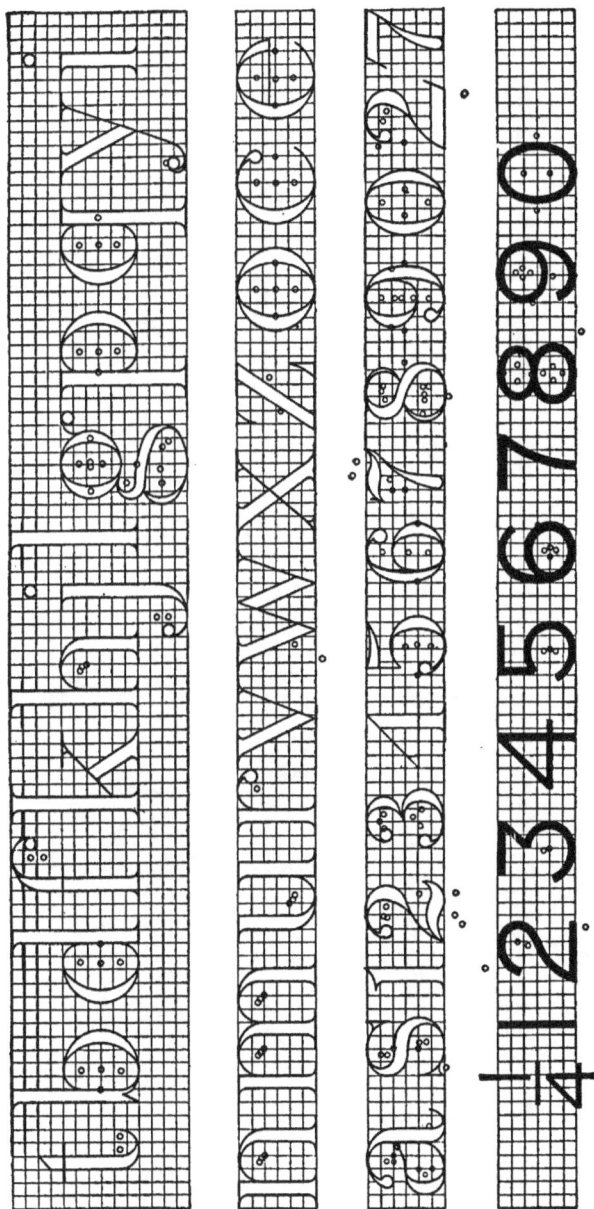

FIG. 174.

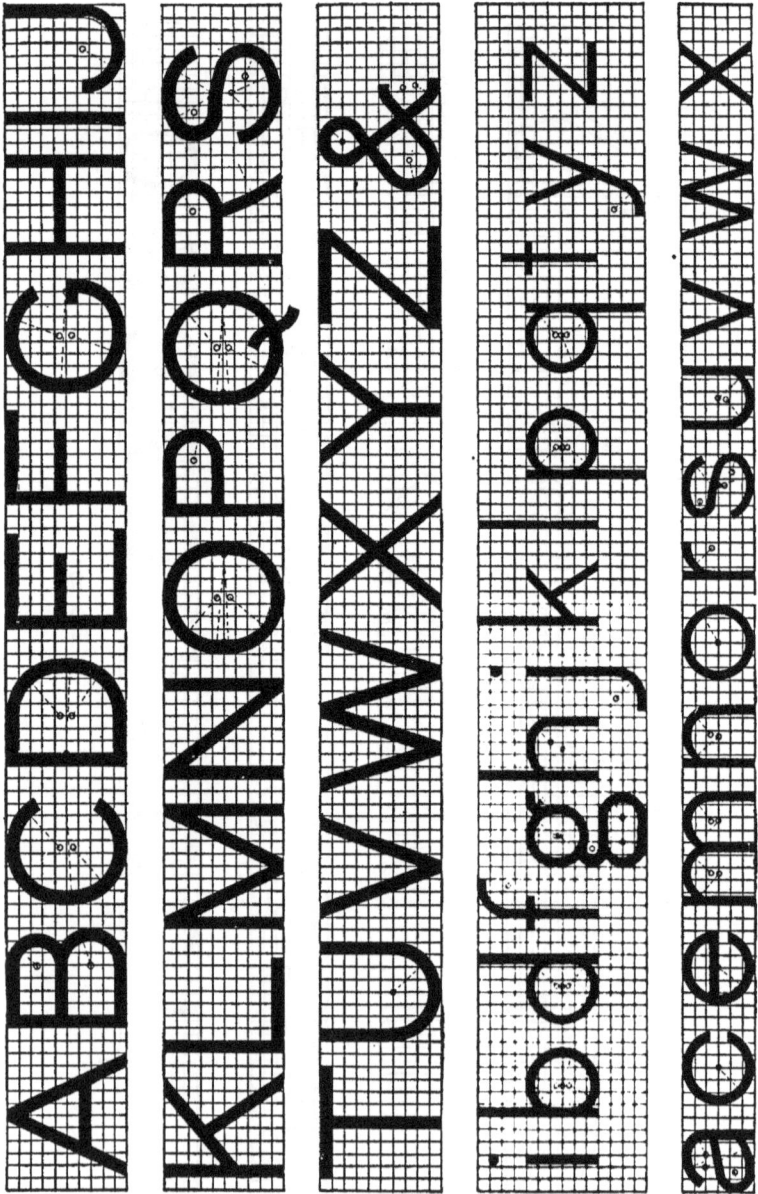

ABCDEFGHIJ
KLMNOPQRS
TUVWXYZ&
bdfghjklpqtyz
acemnorsuvwx

FIG. 175.

The alphabet shown in Figs. 176 and 177, known as the classic Renaissance letters, is selected as a good form of letter for general purposes, where a Roman letter would be suitable for the work in hand.   This alphabet was originally designed by Albrecht

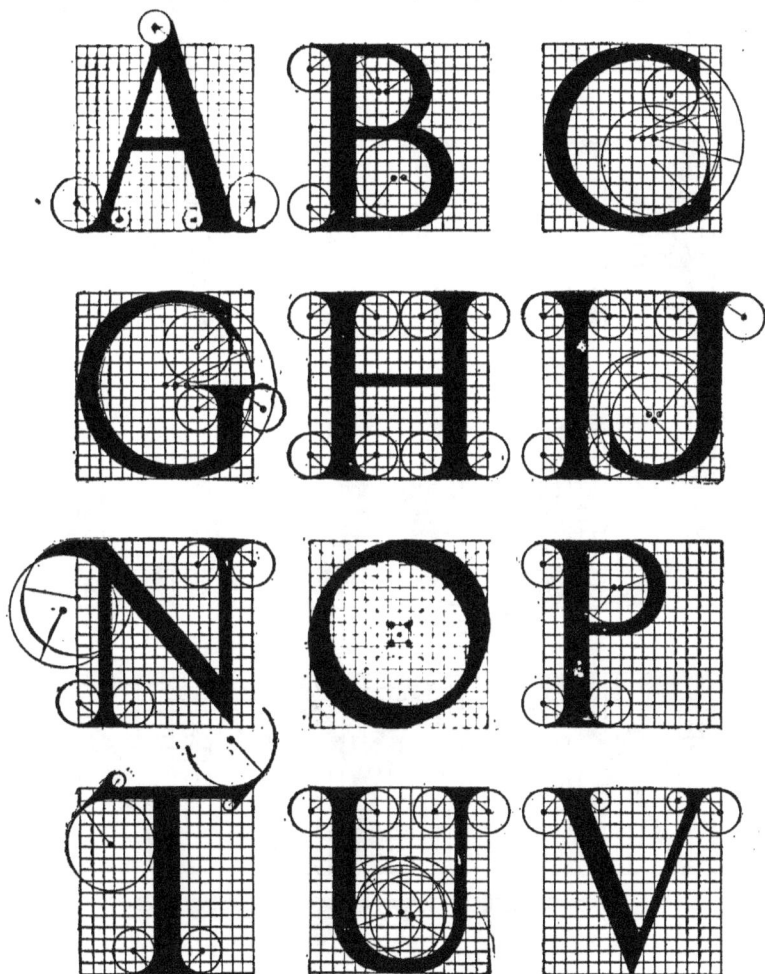

FIG. 176.

Dürer and adopted by Frank Chouteau Brown, in his treatise on " Letters and Lettering," Bates & Guild Company, Boston. Mr. Brown's book is recommended to those students who desire to follow up their studies in architectural lettering.

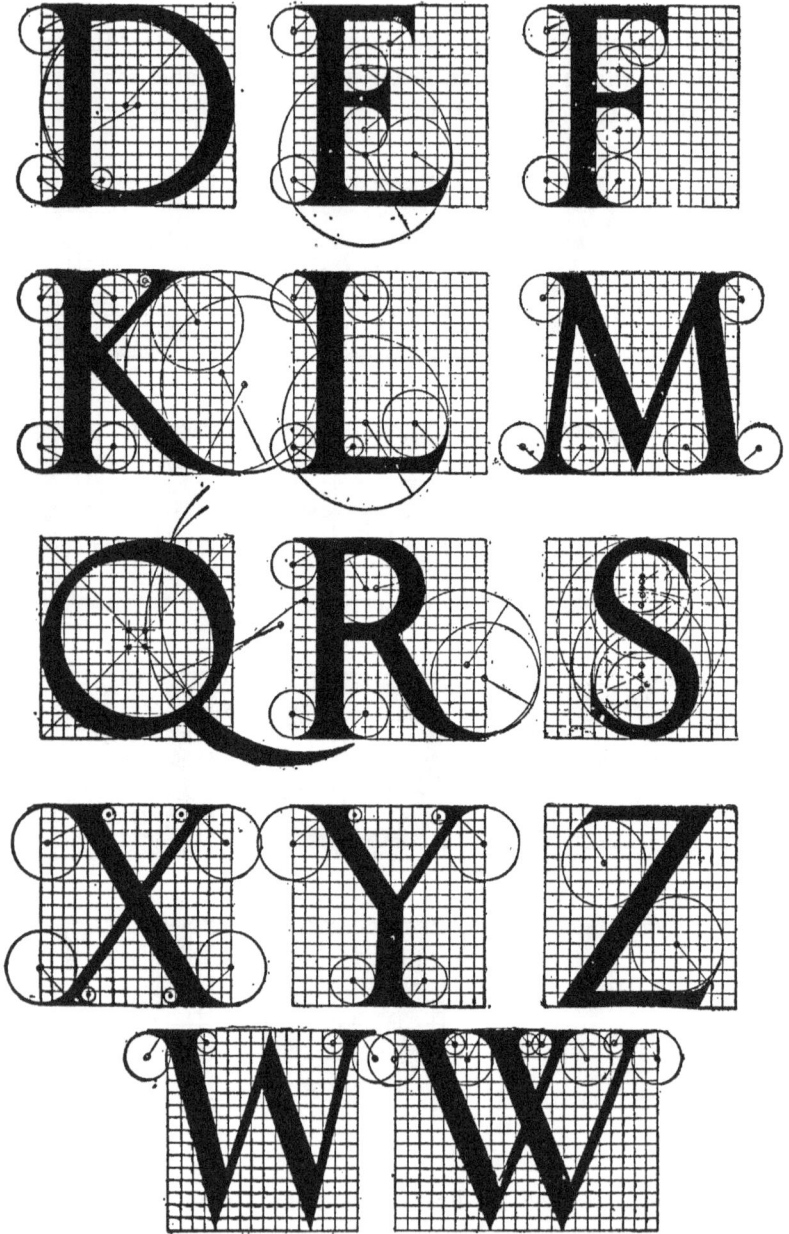

FIG. 177.

18-Point Roman.

ABCDEFGHIJKLMNOPQRSTUVWX
YZ   abcdefghijklmnopqrstuvwxyz
1234567890

18-Point Italic.

ABCDEFGHIJKLMNOPQRSTUV
WXYZ   abcdefghijklmnopqrstuvwxyz

12-Point Cushing Italic.

ABCDEFGHIJKLMNOPQRSTUVWXYZ   abcdefghijklm
nopqrstuvwxyz   1234567890

28-Point Boldface Italic.

ABCDEFGHIJKLM
NOPQRSTUVWXYZ
abcdefghijklmnopqrstu
vwxyz   1234567890

Two-Line Nonpareil Gothic Condensed.

ABCDEFGHIJKLMNOPQRSTUVWXYZ   1234567890

Three-Line Nonpareil Lightface Celtic.

ABCDEFGHIJKLMNOPQR
STUVWXYZ   abcdefghijkl
mnopqrstuvwxyz
1234567890

18-Point Chelsea Circular.

ABCDEFGHIJKLMNOPQRSTUVWX
YZ abcdefghijklmnopqrstuvwxyz
1234567890

18-Point Elandkay.

ABCDEFGHIJKLMNOPQRSTUVWXYZ
1234567890

18-Point Quaint Open.

ABCDEFGHIJKLMNOPQRSTUV
WXYZ 1234567890

28-Point Roman.

ABCDEFGHIJKLM
NOPQRSTUVWXYZ
abcdefghijklmnopqrstu
vwxyz · 1234567890

28-Point Old-Style Italic.

ABCDEFGHIJKLMNOP
QRSTUVWXYZ abcdefg
hijklmnopqrstuvwxyz
1234567890

12-Point Victoria Italic.

ABCDEFGHIJKLMNOPQRSTU
VWXYZ 1234567890

18-Point DeVinne Italic.

ABCDEFGHIJKLMNOPQRSTU
VWXYZ abcdefghijklmnopqrst
uvwxyz 1234567890

22-Point Gothic Italic.

ABCDEFGHIJKLMNOPQRSTUVWXYZ

abcdefghijklmnopqrstuvwxyz

1234567890

Double-Pica Program.

ABCDEFGHIJKLMNO
PQRSTUVWXYZ
abcdefghijklmnopqrstuv
wxyz 1234567890

Nonpareil Telescopic Gothic.

ABCDEFGHIJKLMNOPQRSTUVWXYZ 1234567890

24-Point Gallican.

ABCDEFGHIJKL
MNOPQRSTUVW
XYZ 1234567890

Two-Line Virile Open.

ABCDEFGHIJKLMNOPQRSTUVWXYZ
abcdefghijklmnopqrstuvwxyz
1234567890

22-Point Old-Style Roman.

ABCDEFGHIJKLMNOPQRST
UVWXYZ   abcdefghijklmnopqrst
uvwxyz   1234567890

36-Point Yonkers.

ABCDEFGHIJKLM
NOPQRSTUVWX
YZ abcdefghijklmnopqr
stuvwxyz 1234567890

The method used for the instrumental construction of these.
letters is similar to that used in the Roman letter given on
page 158.

For the purpose of learning the form and proportions of these
letters the alphabet should be drawn mechanically to a scale as
large as convenient; after which practice should be had by forming the letters freehand to small sizes, until the student becomes
familiar with their construction.

In the following pages a number of type specimens are given
that students may have a choice when some special lettering is
desired.

"36-point Yonkers " on page 166 is sometimes used in special
work.   It is easy to construct with F. Soenneckin's round writing-
pens, single point or the automatic shading-pen.   But it lacks
legibility and is therefore not much used.

87. Geometrical Drawing Continued.   Among the following
problems given here, in addition to the regular course, many
of them are of practical value in drafting as well as being good
exercises in drawi g.

FIG. 178. *To Erect a Perpendicular at the End of the Line.*
Assume the point E above the line as center and radius EB,
describe an arc CBD, cutting the line AB in the point C.   From
C draw a line through E, cutting the arc in D.   Draw DB the
perpendicular.

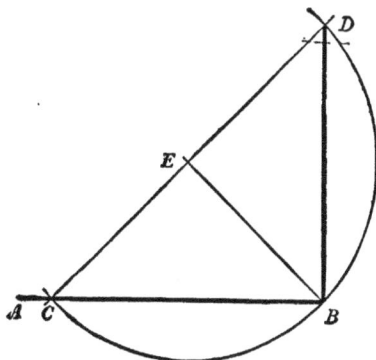

FIG. 178.

FIG. 179. *The Same Problem: A Second Method.*   With
center B and any radius as BC describe an arc CDE with the

same radius; measure off the arcs *CD* and *DE*. With *D* and *E* as centers and any convenient radius describe arcs intersecting at *F*. *FB* is the required perpendicular.

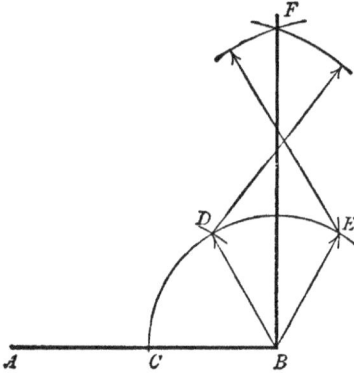 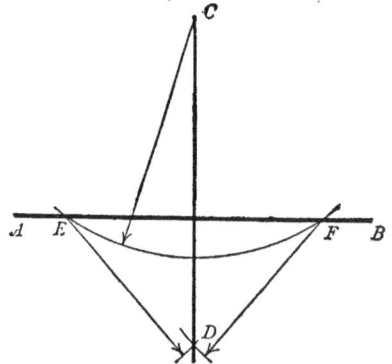

FIG. 179.   FIG. 180.

FIG. 180. *To Draw a Perpendicular to a Line from a Point above or below It.* Assume the point *C* above the line. With center *C* and any suitable radius cut the line *AB* in *E* and *F*. From *E* and *F* describe arcs cutting in *D*. Draw *CD* the perpendicular required.

FIG. 181. *To Draw a Line Parallel to a Given Line AB through a Given Point C.* From any point on *AB* as *B* with radius *BC* describe an arc cutting *AB* in *A*. From *C* with the same radius describe arc *BD*. From *B* with *AC* as radius cut arc *BD* in *D*. Draw *CD*. Line *CD* is parallel to *AB*.

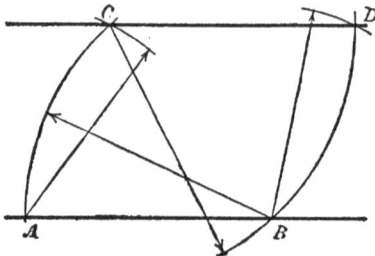

FIG. 181.

FIG. 182. *To Divide a Line AB Proportionally to the Divided Line CD.* Draw *AB* parallel to *CD* at any points 5, 6, 7, 8, etc.

The divisions on AB will have the same proportion to the divisions on CD that the whole line AB has to the whole line CD, i.e., the lines will be proportionally divided.

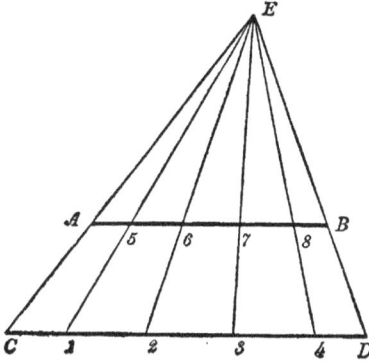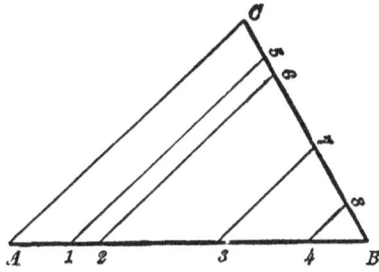

FIG. 182.   FIG. 183.

FIG. 183. *The Same: Another Method.* Let BC, the divided line, make any angle with BA, the line to be divided at B. Draw line CA joining the two ends of the lines. Draw lines from 5, 6, 7, 8, parallel to CA, dividing line AB in points 1, 2, 3, 4, proportional to BC.

FIG. 184. *To Construct a Square, its Base AB being Given.* Erect a perpendicular at B. Make BC equal to AB. From A and C with radius AB describe arcs cutting in D. Join DC and DA.

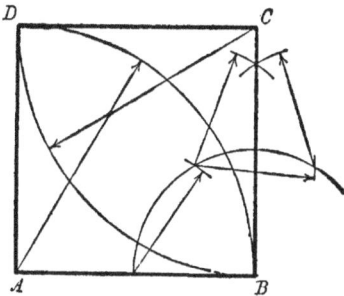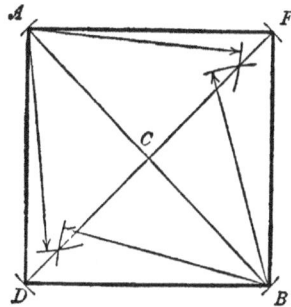

FIG. 184.   FIG. 185.

FIG. 185. *To Construct a Square Given Its Diagonal AB.* Bisect AB in C. Draw DF perpendicular to AB at C. Make CD and CF each equal to CA. Join AD, DB, BF and FA.

FIG. 186. *To Construct a Regular Polygon of Any Number of Sides given the Circumscribing Circle.* Draw a diameter AB of the given circle. Divide AB into as many equal parts as the polygon is to have sides, say 5. From A and B with the line AB as radius describe arcs cutting in C, draw a line from C through the second division of the diameter and produce it cutting the circle in D. BD will be the side of the required polygon. The line C must always be drawn through the second division of the diameter, whatever the number of sides of the polygon.

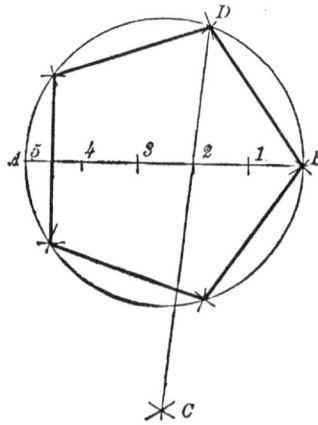

FIG. 186.

FIG. 187. *To Construct any Regular Polygon with a Given Side* AB. Make BD perpendicular and equal to AB. With B as center and radius AB describe arc DA. Divide arc DA into as many equal parts as there are sides in the required polygon, as 1, 2, 3, 4, 5. Draw $B2$. Bisect line AB and erect a perpendicular at the bisection, cutting B 2in C. With C as center and radius CB describe a circle. With AB as a chord step off the remaining sides of the polygon.

FIG. 188. *Another Method.* Extend line AB. With center A and any convenient radius describe a semicircle. Divide the semicircle into as many equal parts as there are sides in the required polygon, say 6. Draw lines through every division except the first. With A as center and AB as radius cut off $A2$

in *C*. From *C* with the same radius cut A_3 in *D*. From *D*, A_4 in *E*. From *B*, A_5 in *F*. Join *AC*, *CD*, *DE*, *EF*, and *FB*.

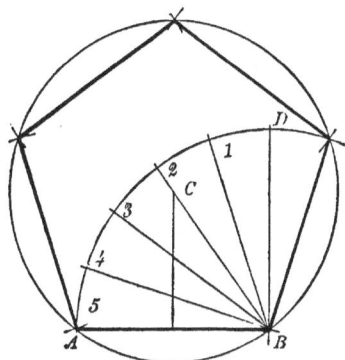

FIG. 187.

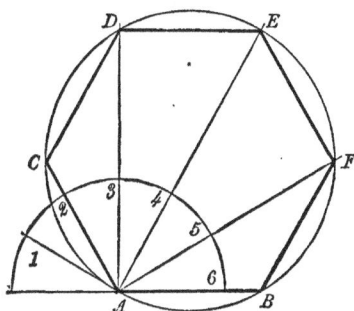

FIG. 188.

FIG. 189. *To Construct a Regular Heptagon, the Circumscribing Circle being Given.* Draw a radius *AB*. With *B* as center and *BA* as radius, cut the circumference in 1, 2; it will be bisected by the radius in *C*. C_1 or C_2 is equal to the side of the required heptagon.

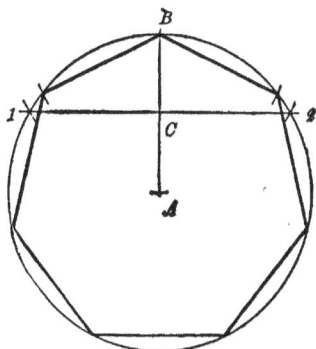

FIG. 189.

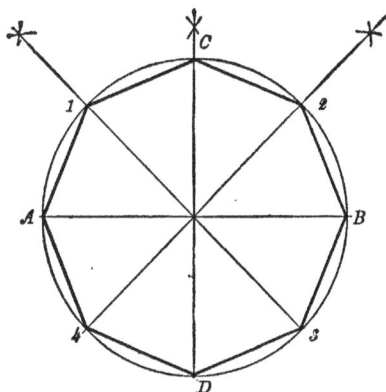

FIG. 190.

FIG. 190. *To Construct a Regular Octagon, the Circumscribing Circle being Given.* Draw a diameter *AB*. Bisect the arcs *AB* in *C* and *D*. Bisect arcs *CA* and *CB* in 1 and 2. Draw lines from 1 and 2 through the center of the circle, cutting the circumference in 3 and 4. Join A1, 1C, C2, 2B, B3, etc.

FIG. 191. *To Inscribe an Octagon in a Given Square.* Draw
diagonals AD, CB intersecting at O. From A, B, C, and D with
radius equal to AO describe quadrants cutting the sides of the
square in 1, 2, 3, 4, 5, 6, 7, 8. Join these points and the octagon
will be inscribed.

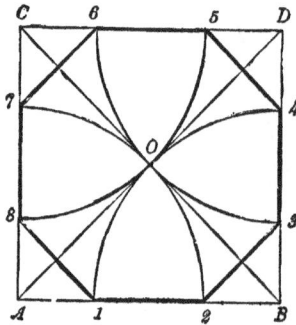

FIG. 191.

FIG. 192. *To Find a Fourth Proportional to Three Given Right
Lines AB, CD, and EF.* Make $GH =$ the given line AB. Draw
$GI = CD$, making any convenient angle to GH. Join HI. From
G lay off $GK = EF$. From K draw a parallel to HI cutting GI in
L. GL is the fourth proportional required.

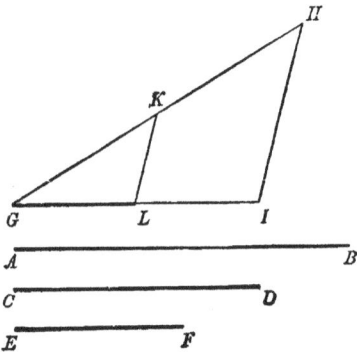

FIG. 192.

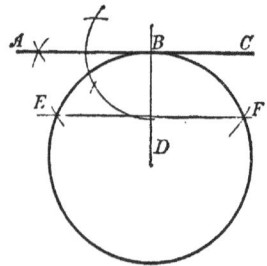

FIG. 193.

FIG. 193. *To Draw a Line Tangent to an Arc of a Circle.*
(1st) When the center is not accessible. Let B be the point

through which the tangent is to be drawn.   From *B* lay off equal
distances as *BE, BF*.   Join *EF* and through *B* draw *ABC* parallel to *EF*.   (2d) When the center *D* is given.   Draw *BD* and
through *B* draw *ABC* perpendicular to *BD*.   *ABC* is tangent
to the circle at the point *B*.

FIG. 194. *To Draw Tangents to the Circle C from the Point A
Without It*.   Draw *AC* and bisect it in *E*.   From *E* with radius
EC describe an arc cutting circle *C* in *B* and *D*.   Join *CB, CD*.
Draw *AB* and *AD* tangent to the circle *C*.

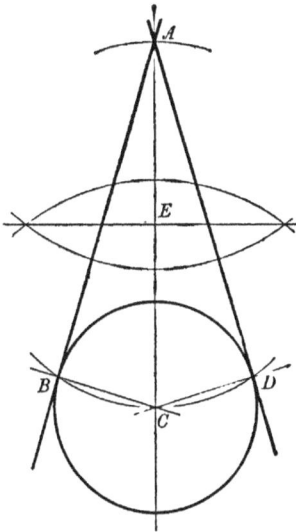

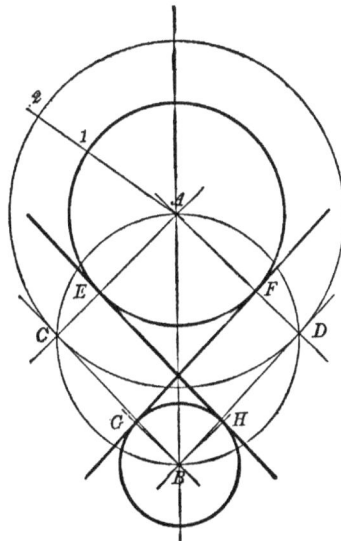

FIG. 194.   FIG. 195.

FIG. 195. *To Draw a Tangent between Two Circles*.   Join the
centers *A* and *B*.   Draw any radial line from *A* as *A*2 and make
1–2 = the radius of circle *B*.   From *A* with radius *A*–2 describe a
circle *C*2*D*.   From center *B* draw tangents *BC* and *BD* to circle
*C*2*D* at the points *C* and *D* by preceding problem.   Jcin *AC* and
AD and through the points *E* and *F* draw parallels *FG* and *EH*
to *BD* and *BC*.   *FG* and *EH* are the tangents required.

FIG. 196. *To Draw Tangents to Two Given Circles A and B*.
Join *A* and *B*.   From *A* with a radius equal to the difference of
the radii of the given circles describe a circle *CF*.   From *B* draw

the tangents *BF* and *BG*, by Prob. 37. Draw *AF* and *AG* extended to *E* and *H*. Through *E* and *H* draw *EC* and *HD* parallel to *BF* and *BG* respectively. *EC* and *DH* are the tangents required.

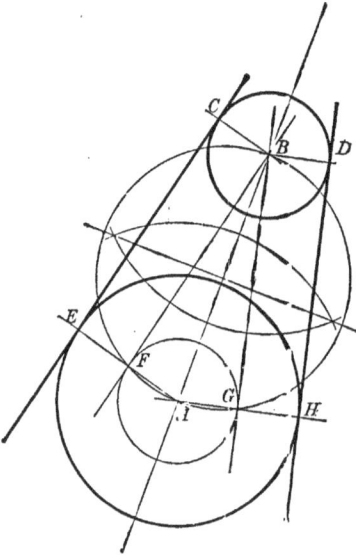

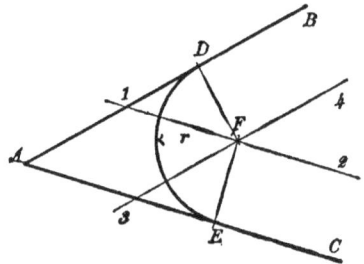

FIG. 196.   FIG. 197.

FIG. 197. *To Draw an Arc of a Circle of Given Radius Tangent to Two Straight Lines.* *AB* and *AC* are the two straight lines, and *r* the given radius. At a distance=*r* draw parallels 1–2 and

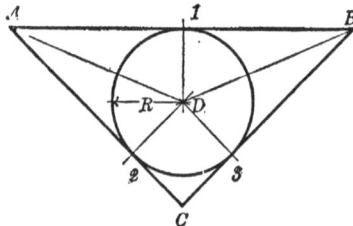

FIG. 198.

3–4 to *AC* and *AB*, intersecting at *F*. From *F* draw perpendiculars *FD* and *FE*. With *F* as center and *FD* or *FE* as radius describe the required arc, which will be tangent to the two straight lines at the points *D* and *E*.

FIG. 198. *To Inscribe a Circle within a Triangle ABC.* Bisect the angles A and B. The bisectors will meet in D. Draw $D1$ perpendicular to AB. Then with center D and radius $= D1$ describe a circle which will be tangent to the given triangle at the points 1, 2, 3.

FIG. 199. *To Draw an Arc of a Circle of Given Radius R Tangent to Two Given Circles A and B when the Arc Includes One Circle and Excludes the Other.* Through A draw any diameter and make $1-2 = R$. From B draw any radius and extend it, making $3-4 = R$.

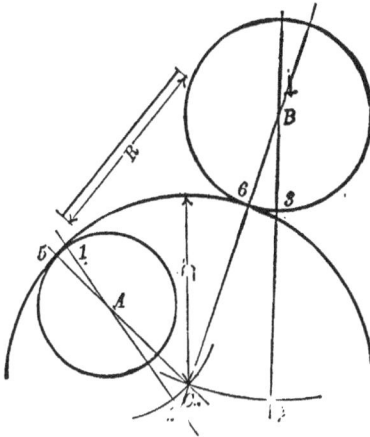

FIG. 199.

With center A and radius $A2$ and center B and radius $B4$ describe arcs cutting at C. With C as center and radius $= C5$ or $C6$ describe the arc 5, 6.

FIG. 200. *Draw an Arc of a Circle of Given Radius R Tangent to a Straight Line AB and a Circle CD.* From E, the center of the given circle, draw an arc of a circle 1, 2 concentric with CD at a distance R from it, and also a straight line 3, 4 parallel to AB at the same distance R from AB. Draw EO intersecting CD at 5. Draw the perpendicular $O6$. With center O and radius $O6$ or $O5$ describe the required arc.

FIG. 201. *To Describe an Ellipse Approximately by Means of Three Radii.* (F. R. Honey's method.) Draw straight lines

RH and HQ, making any convenient angle at H. With center H
and radii equal to the semi-minor and semi-major axes respectively, describe arcs LM and NO. Join LO and draw MK and
NP parallel to LO. Lay off $L1 = \frac{1}{8}$ of LN. Join $O1$ and draw

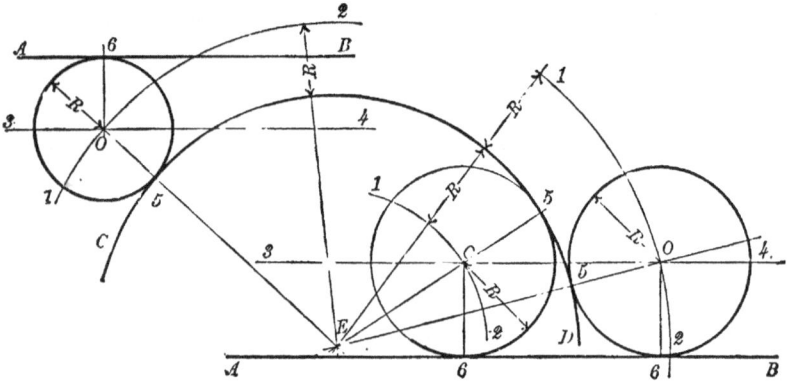

FIG. 200.

$M2$ and $N3$ parallel to $O1$. Take $H3$ for the longest radius
$(=T)$, $H2$ for the shortest radius $(=E)$, and one-half the sum
of the semi-axes for the third radius $(=S)$, and use these radii to
describe the ellipse as follows: Let AB and CD be the major

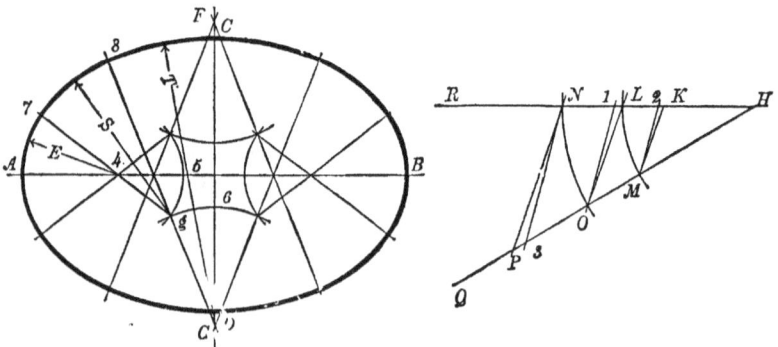

FIG. 201.

and minor axes. Lay off $A4 = E$ and $A5 = S$. Then lay off CG
$=T$ and $C6 = S$. With G as center and $G6$ as radius draw the
arc 6, g. With center 4 and radius 4, 5, draw arc 5, g, intersecting 6, g at g. Draw the line Gg and produce it making $G8 = T$.

Draw g, 4 and extend it to 7 making g, $7 = S$. With center G and radius $GC(=T)$ draw the arc $C8$. With center g and radius g, 8 ($=S$) draw the arc 8, 7. With center 4 and radius 4, 7 ($=E$) draw arc $7A$. The remaining quadrants can be drawn in the same way.

FIG. 202. *To Describe an Ellipse Given the Semi-axes AB and CD.* Let AB and AC be the semi-axes. With A as center and radii AB and AC describe circles. Draw any radii as $A3$ and $A4$, etc. Make 3 1, 4 2, etc., perpendicular to AB, and $D2$, $E5$, etc., parallel to AB. Then 1, 2, 5, etc., are points on the curve.

FIG. 203. *Another Method.* Place the diameters as before, and construct the rectangle $CDEF$. Divide AB and DB and BF into the same number of equal parts as 1, 2, 3, and B. Draw

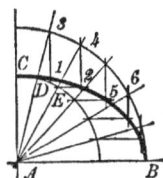

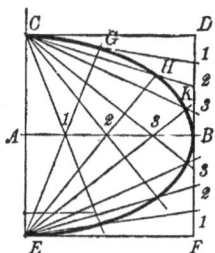

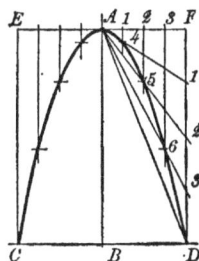

FIG. 202.   FIG. 203.   FIG. 204.

from C through points 1, 2, 3 on AB and BD lines to meet others drawn from E through points 1, 2, 3 on AB and FB intersecting in points GHK. GHK are points on the curve.

FIG. 204. *To Construct a Parabola, the Base CD and the Abscissa AB Being Given.* Draw EF through A parallel to CD and CE and DF parallel to AB. Divide AE, AF, EC, and FD into the same number of equal parts. Through the points 1, 2, 3 on AF and AE draw lines parallel to AV, and through A draw lines to the points 1, 2, 3 on FD and EC intersecting the parallel lines in points 4, 5, 6, etc., of the curve.

FIG. 205. *Given an Ellipse to Find the Axes and Foci.* Draw two parallel chords AB and CD. Bisect each of these in E and F. Draw EF touching the ellipse in 1 and 2. This line divides the ellipse obliquely into equal parts. Bisect 1, 2 in G, which will

be the center of the ellipse.   From G with any radius draw a
circle cutting the ellipse in $HIJK$.   Join these four points and a
rectangle will be formed in the ellipse.   Lines LM and NO,
bisecting the sides of the rectangle, will be the diameters or axes
of the ellipse.   With N or O as centers and radius $= GL$ the semi-
major axis, describe arcs cutting the major axis in P and Q the
foci.

  FIG. 206. *To Construct a Spiral of One Revolution.*   Describe
a circle using the widest limit of the spiral as a radius.   Divide

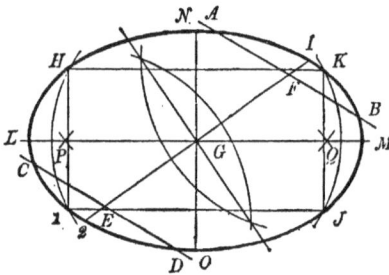

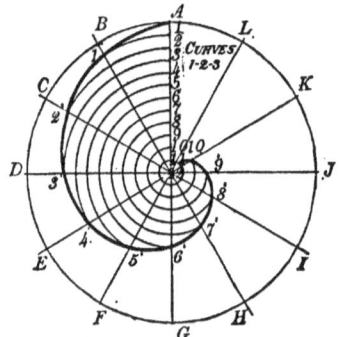

FIG. 205.   FIG. 206.

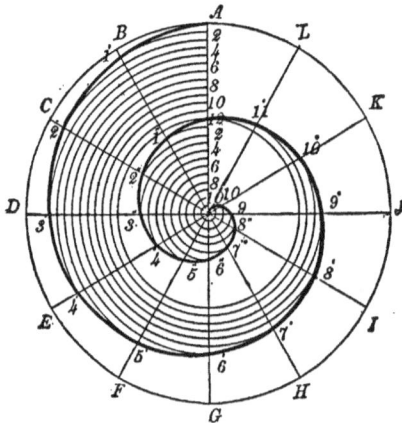

FIG. 207.

the circle into any number of equal parts as A, B, C, etc.   **Divide**
the radius into the same number of equal parts as 1 to 12.   From
the center with radius 12, 1 describe an arc cutting the radial
line B in 1'.   From the center continue to draw arcs from points

2, 3, 4, etc., cutting the corresponding radii C, D, E, etc., in the points 2′, 3′, 4′, etc. From 12 trace the Archimedes Spiral of one revolution.

Fig. 207. *To Describe a Spiral of any Number of Revolutions*, *e.g.*, 2. Divide the circle into any number of equal parts as A, B, C, etc., and draw radii. Divide the radius $A12$ into a number of equal parts corresponding with the required number of revolutions and divide these into the same number of equal parts as there are radii, viz., 1 to 12. It will be evident that the figure consists of two separate spirals, one from the center of the circle to 12, and one from 12 to A. Commence as in the last problem, drawing arcs from 1, 2, 3, etc., to the correspondingly numbered radii, thus obtaining the points marked 1′, 2′, 3′, etc. The first revolution completed, proceed in the same manner to find the points 1″, 2″, 3″, etc. Through these points trace the spiral of two revolutions.

Fig. 208. *To Describe the Cycloid.* AB is the director, CB the generating circle, X a piece of thin transparent celluloid, with

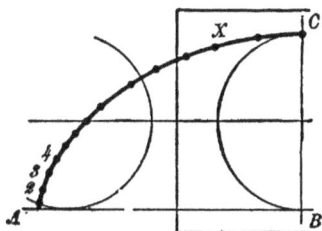

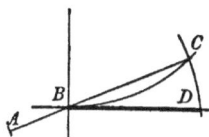

FIG. 208.   FIG. 209.

one side dull on which to draw the circle C. At any point on the circle C puncture a small hole with a sharp needle, and place the point C tangent to the director AB at the point from which the curve is to be drawn. Hold the celluloid at this point with a needle, and rotate it until the arc of the circle C intersects the director AB. Through the point of intersection stick another needle and rotate X until the circle is again tangent to AB, and through the puncture at C with a 4H pencil, sharpened to a fine conical point, mark the first point on the curve. So proceed until sufficient points have been found to complete the curve.

(NOTE.—The thin celluloid was first used as a drawing instrument by Professor H. D. Williams, of Sibley College, Cornell University.)

FIG. 209. *To Find the Length of a Given Arc of a Circle Approximately.* Let *BC* be the given arc. Draw its chord and produce it to *A*, making *BA* equal half the chord. With center *A* and radius *AC* describe arc *CD* cutting the tangent line *BD* at *D*, and making it equal to the arc *BC*.

FIG. 210. *To Describe the Cycloid by the Old Method.* Divide the director and the generating circle into the same number of

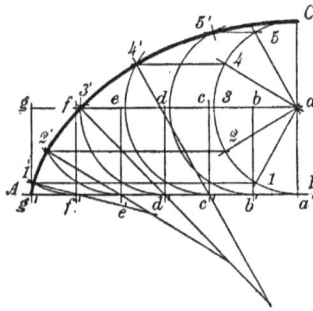

FIG. 210.

equal parts. Through the center *a* draw *ag* parallel to *AB* for the line of centers, and divide it as *AB* in the points *b*, *c*, *d*, *e*, *f*, and *g*. With centers *f*, *e*, *d*, etc., describe arcs tangent to *AB*, and through the points of division on the generating circle 1, 2, 3, etc., draw lines parallel to *AB* cutting the arcs in the points 1′, 2′, 3′, etc. These will be points in the curve.

An approximate curve may be drawn by arcs of circles. Thus, taking *f′* as center and *f′g′* as radius, draw arc *g′1′*. Produce 1′*f′* and 2′*e′* until they meet at the center of the second arc 2′*f′*, etc.

FIG. 211. *Another Method.* Draw the generating circle on the celluloid and roll it on the outside of the director *BC* for the Epicycloid, and on the inside for the Hypocycloid.

FIG. 212. *To Draw the Cissoid.* Draw any line *AB* and *BC* perpendicular to it. On *BC* describe a circle. From the extrem-

88888888888888887888888888888888888888888888

ity C of the diameter draw any number of lines, at any distance apart, passing through the circle and meeting the line AB in $1'$, $2'$, $3'$, etc. Take the length from A to 9 and set it off from C on the same line to $9''$. Take the distance from $8'$ and set it off from C on the same line to $8''$, etc., for the other divisions, and through $9''$, $8''$, $7''$, $6''$, etc., draw the curve.

FIG. 211.

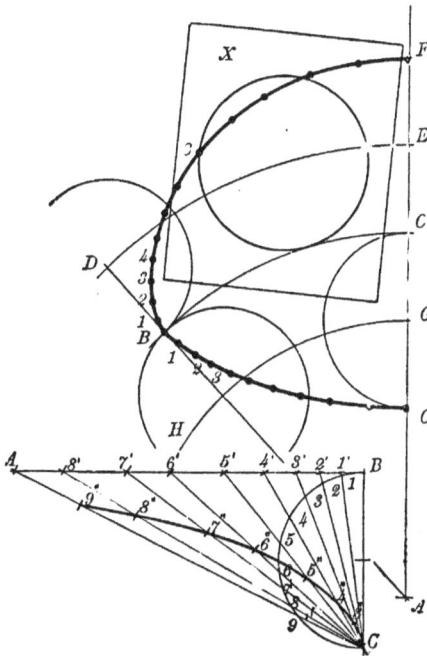

FIG. 212.

FIG. 213. *To Draw Schiele's Anti-friction Curve.* Let AB be the radius of the shaft and $B1$, 2, 3, 4, etc., its axis. Set off the radius AB on the straight edge of a piece of stiff paper or thin celluloid and placing the point B on the division 1, of the axis, draw through point A the line $A1$. Then lower the straight edge until the point B coincides with 2 and the point A just touches the last line drawn, and draw $a2$, and so proceed to find the points a, b, c, etc. Through these points draw the curve.

FIG. 214. *To Describe an Interior Epicycloid.*   Let the large circle X be the generator and the small circle Y the director. Divide circle Y into any number of equal parts, as B, H, I, J, etc. Draw radial lines and make HC, ID, JE, KF, etc., each equal to the radius of the generator X. With centers C, D, E, etc., describe arcs tangent at H, I, J, etc.   Make $H1$ equal to one of the divisions of the director as BH.   Make $I2$ equal to two divisions, $J3$, three divisions, etc., and draw the curve through the points 1, 2, 3, 4.

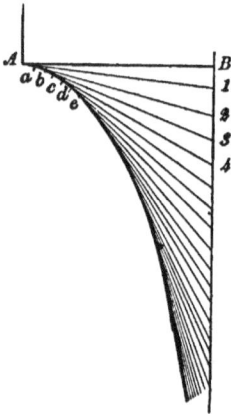

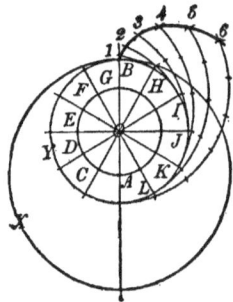

FIG. 213.   FIG. 214.

FIG. 215. *To Construct a Scale of Sixth Size or* $2''=1$ *Foot.* Draw upon a piece of tough white drawing-paper two parallel lines about $1''$ apart and about $14''$ long as shown by a, Fig. 98. From A lay off distances equal to $2''$ and divide the first space AB into 12 equal parts or inches by Prob. 12.   Divide AE in the same way into as many parts as it may be desired to subdivide the inch divisions on AB, usually 8.   When the divisions and subdivisions have been carefully and lightly drawn in pencil as shown by a, in Fig. 215, then the lines denoting $\frac{1}{8}'', \frac{1}{4}'', \frac{1}{2}'', 1''$, and $3''$ should be carefully inked and numbered as shown by (b). By a further subdivision a scale of $1''=1$ foot may easily be made as shown by (c) in Fig. 215.

FIGS. 216 and 217.   Draw the projections of a circular plane (1) when its surface is parallel to the vertical plane, (2) when it

makes an angle of 45° with the V.P., and (3) when still making
an angle of 45° with the V.P. it has been revolved through an
angle of 60°.

First position: Draw the circular plane 1^v, 2^v, 3^v, 4^v, etc.,

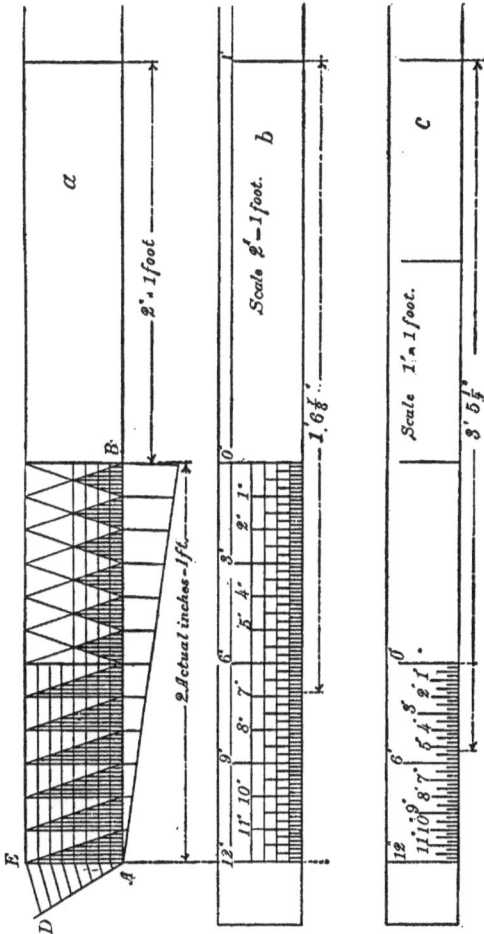

Fig. 215.

Fig. 216, below the I.L. with a radius = $1\frac{1}{2}''$ and divide and
figure it as shown.

Since the plane is parallel to V.P. its horizontal projection
will be a straight line 1^h, 2^h, ... etc.

For the second position revolve the said horizontal projection
through the required angle of 45° to the position a^h . . . . $11_1{}^h$,

Fig. 216, and through each division in 1^h . . . . a^h draw arcs
cutting a^h . . . . 1^h in points 2^h3^h . . . . This is the horizontal projection of the plane when making an angle of 45°
with the V.P.

The elevation is found by dropping perpendiculars from the
points in the horizontal projections a^h . . . 11^h to intersect
horizontal lines drawn through the correspondingly numbered

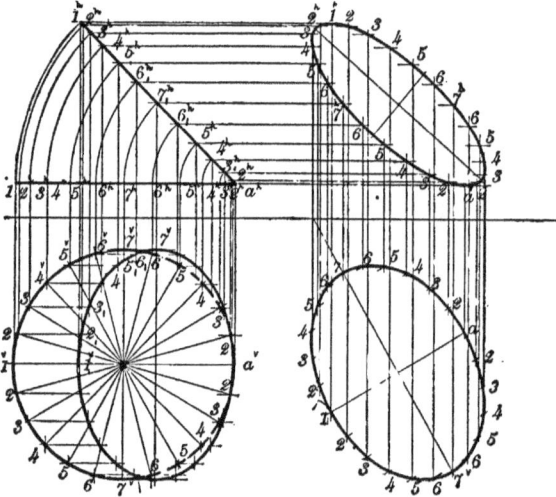

FIG. 216.   FIG. 217.

points in the elvation and through these intersections draw the
elevation or vertical project of the second position.

For the third position make a tracing of the elevation of the
second position, numbering all the points as before, and place
the tracing so that the diameter 7^v7^v makes the required angle of
60° with the I.L. and transfer to the drawing-paper, Fig. 217.

FIG. 218. *Draw the projections of a regular hexagonal prism,
3″ high and having an inscribed circle of $4\frac{5}{8}″$ diameter:* (1)
When its axis is parallel to the V.P. (2) *Draw the true form of
a section of the prism when cut by a plane passing through it a
an angle of 30° with its base.* (3) *Draw the projection of a section when cut by a plane passing through XX, Fig. 218, perpendicular to both planes of projection.*

The drawing of the I.L. may now be omitted.

For the plan of the first part of this problem draw a circle with a radius = to $2\frac{5}{16}''$, and circumscribe a hexagon about it, as shown by a^h, b^h, b^h, etc., Fig. 218. To project the elevation, draw at a convenient distance from the plan a horizontal line parallel to $a^h d^h$, and $3''$ below it another line parallel to it. From the

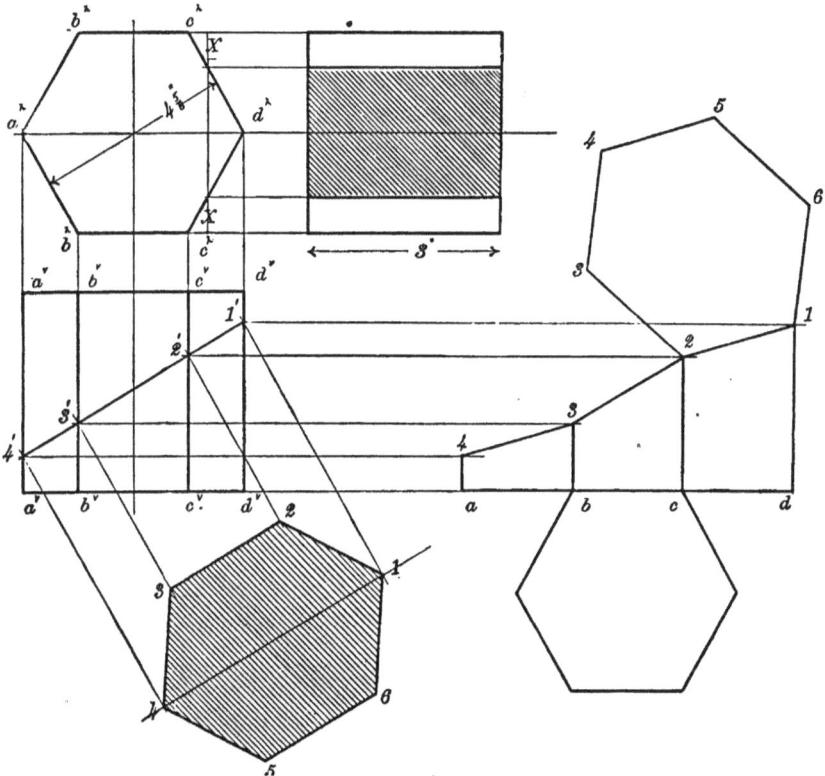

FIG. 218.   FIG. 219.

points $a^h b^h c^h d^h$, drop perpendiculars cutting these parallel lines in the points $a^v b^v c^v d^v$, thus completing the elevation of the prism.

Second condition: Draw the edge view or trace of the cutting plane $1'4'$, making an angle of 30° with the base of the prism, locating the lower end $4'$ $\frac{1}{2}''$ above the base; parallel to $1'4'$, and at a convenient distance from it draw a straight line 1, 4; at a distance of $2\frac{5}{16}''$ on each side of 1, 4 draw lines 3, 2 and 5, 6 parallel to 1, 4, and through the points $1'2'3'4'$ let fall perpendiculars cutting these three parallel lines in the points 1, 2, 3, 4, 5, 6; join these points by straight lines as shown, and a true drawing of the section of the prism as required will result.

For the third condition of the problem:

Let XX be the edge view of the cutting plane and conceive that part of the prism to the right of XX to be removed. From the horizontal projection of the prism draw a right-hand elevation or profile projection, and through the points XX draw the lines enclosing the section, and hatch-line it as shown.

FIG. 219. *To draw the development of the lower part of the prism in the elevation of the last problem.*

To the right of the elvation in Fig. 218, prolong the base line indefinitely and lay off upon it the distances ab, bc, cd, etc., Fig. 216, each equal in length to a side of the hexagon. At these points erect perpendiculars, and through the points $1'2'3'4'$ draw horizontal lines intersecting the perpendiculars in 4, 3, 2, 1, etc. At be draw the hexagon $a^h b^h b^h$, $c^h c^h$, d^h of the last problem for the base, and at 1, 2 draw the section 1, 2, 3, 4, 5, 6 for the top.

FIG. 220. *To draw the projections of a right cylinder 3' diameter and 3″ long. (1) When its axis is perpendicular to the H.P. (2) Draw the true form of a section of the cylinder, when cut by a plane perpendicular to the V.P., making an angle of 30° with the H.P. (3) Draw the development of the upper part of the cylinder.*

For the plan of the first condition, describe the circle $1'$, $2'$, etc., with radius $= 1\frac{1}{2}''$ and from it project the elevation, which will be a square of $3''$ sides.

For the second condition: Let 1, 7 be the trace of the cutting plane, making the point 7, $\frac{1}{2}''$ from the top of the cylinder. Divide the circle into 12 equal parts and let fall perpendiculars through these divisions to the line of section, cutting it in the points 1, 2, 3, 4, etc. Parallel to the line of Section 1, 7 draw $1''7''$ at a convenient distance from it, and through the points 1, 2, 3, 4, etc., draw perpendiculars to 1, 7, intersecting and extending beyond $1''7''$. Lay off on these perpendiculars the distances $6''8'' = 6'8'$, and $5''9'' = 5'9'$, etc., and through the points $2''$, $3''$, $4''$, etc., describe the ellipse.

For the development: In line with the top of the elevation
draw the line $g'g''$ equal in length to the circumference of the
circle, and divide it into 12 equal parts a', b', etc., a', b'', etc.
Through these points drop perpendiculars and through the points
1, 2, 3, etc., draw horizontals intersecting the perpendiculars in
the points 1, 2, 3, etc., and through these points draw a curve.

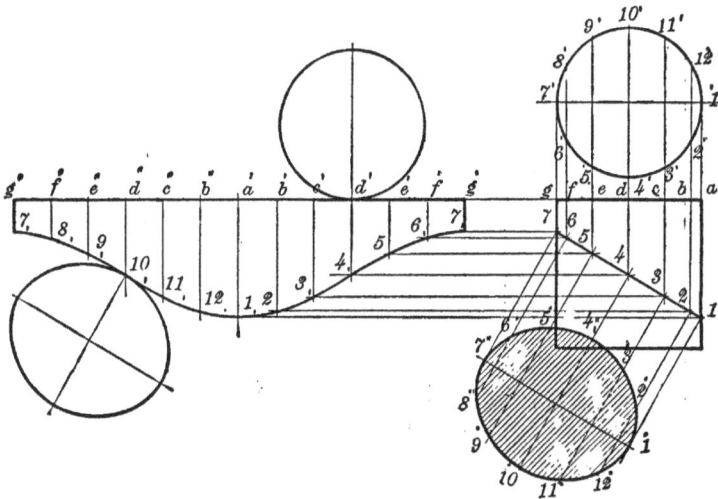

FIG. 220.

Tangent to any point on the straight line draw a 3″ circle
for the top of the cylinder and tangent to any suitable point on
the curve transfer a tracing of the ellipse.

FIG. 221. *Draw the projections of a right cone 7″ high,
with a base 6″ in diam., pierced by a right cylinder 2″ in diameter and 5″ long, their axes intersecting at right angles 3″ above
the base of the cone and parallel to V.P. Draw first the plan of
the cone with a radius* $=3''$.

At a convenient distance below the plan draw the elevation
to the dimensions required.

3″ above the base of the cone draw the center line of the
cylinder *CD*, and about it construct the elevation of the cylinder,
which will appear as a rectangle 2″ wide and $2\frac{1}{2}$″ each side of the
axis of the cone. The half only appears in the figure.

To project the curves of itersection between the cylinder

and cone in the plan and elevation: Draw to the right of the cylinder on the same center line a semicircle with a radius equal that of the cylinder.   Divide the semicircle into any number of parts, as 1, 2, 3, 4, etc.   Through 1, 1 draw the perpendiculars A'' 1$''$ equal in length to the height of the cone, and through A'' draw the line $A''4''$ tangent to the semicircle at the point 4, and

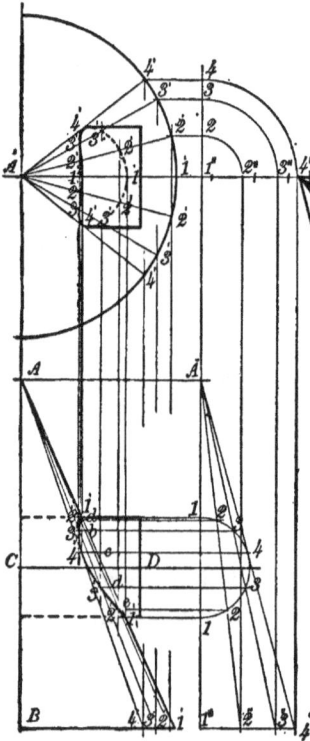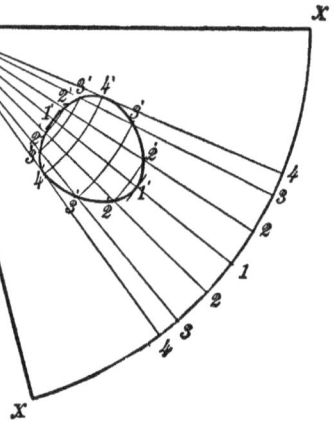

FIG. 221.   FIG. 222.

through the other divisions of the semicircle draw lines from A'' to the line 1$''$4$''$, meeting it in the points 3$''$2$''$.

From all points on the line 1$''$4$''$, viz., 1$''$2$''$3$''$4$''$, erect perpendiculars to the center line of the plan, cutting it in the points 1$_1$$''4_1$$''3_1$$''4_1$$''$, and with 1$_1$$''$ as the center draw the arcs 2$_1$$''$-2, 3$_1$$''$-3, 4$_1$$''$-4 above the center line of the plan, and through the points 2, 3, 4 draw horizontals to intersect the circle of the plan in the points 2$'$3$'$4$'$, and lay off the same distances on the other

side of the center line of the plan in same order, viz., $2'3'4'$.
Through each of these points on the circumference of the circle
of the plan draw radii to its center A', and through the same
points also in the plan let fall perpendiculars to the base of the
elevation of the cone, cutting it in the points $2'3'4'$; and from the
apex A of the elevation of the cone draw lines to the points $2'3'4'$
on the base. Horizontal lines drawn through the points of
division 2, 3, 4 on the semicircle will intersect the elements A–$2'$,
A–$3'$, A–$4'$ of the cone in the points $2'3'4'$; these will be points in
the elevation of the curve of intersection between the cylinder
and the cone.   .

The plan of the curve is found by erecting perpendiculars
through the points in the elevation of the curve to intersect the
radial lines of the plan in correspondingly figured points, through
which trace the curve as shown. Repeat for the other half of the
curve.

FIG. 222. To draw the development of the half cone, showing the hole penetrated by the cylinder.

With center $41''$, Fig. 222, and element $A1'$ of the cone,
Fig. 128, as radius, describe an arc equal in length to the. semicircle of the base of the cone. Bisect it in the line $41''1$, and
on each side of the point 1 lay off the distances 2, 3, 4, equal
to the divisions of the arc in the plan Fig. 128, and from these
points draw lines to $4''$, the center of the arc. Then with radii
A–a, b, c, d, e, respectively, on the elevation Fig. 128, and center
$41''$ draw arcs intersecting the lines drawn from the arc XX to
its center $41''$. Through the points of intersection draw the
curve as shown by Fig. 222.

FIG. 223. *To draw the development of the half of a truncated cone, given the plan and elevation of the cone.*

Divide the semicircle of the plan into any number of parts,
then with A as center and $A1$ as radius, draw an arc and lay
off upon it from the point 1 the divisions of the semicircle from
1 to 9, draw $9A$. Then with center A and radius AB draw the
arc BC. $1BC9$ is the development of the half of the cone approximately.

FIG. 224. *To draw the curve of intersection of a small*

cylinder with a larger. To the left of the center-line of **Fig.**
224 is a half cross-section, and to the right a half elevation
of the two cylinders.

Draw the half plan of the small cylinder, which will be a
semicircle, and divide it into any convenient number of parts,
say 12.

From each of these divisions drop perpendiculars.

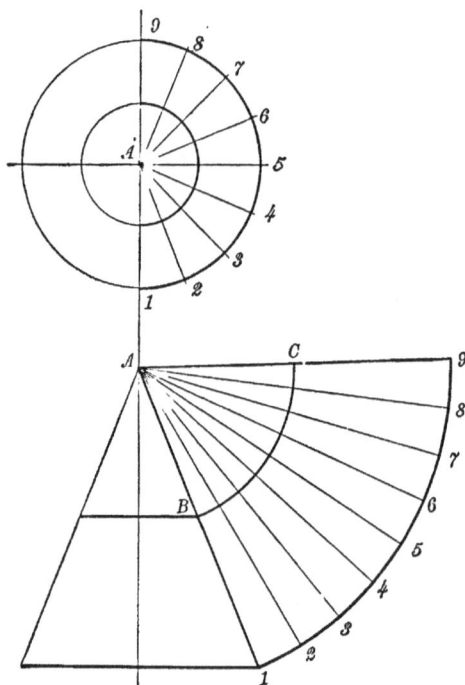

Fig. 223.

On the half cross-section these perpendiculars intersect
the circumference of the large cylinder in the points 1′, 2′, etc.
Through these points draw horizontals to intersect in corresponding points the perpendiculars on the half elevation.
Through the latter points draw the curve of intersection *C*.

*To draw the development of the smaller cylinder of the last
problem.*

Draw a rectangle, Fig. 225, with sides equal to the circumference and length of the cylinder, respectively, and divide it into 24 equal parts.

Make AB'', 1 $1'$, 3 $3'$, etc., Fig. 225, equal to AB, $1'1''$, $2'2''$, $3'3''$, etc., Fig. 224, and draw the developed curve of intersectio1.

To draw the orthographic projections of a cylindrical dome riveted to a cylindrical boiler of given dimensions.

Let the dimensions of the dome and boiler be: dome $26\frac{1}{2}''$ diameter $\times 27''$ high, boiler $54''$ diameter, plates $\frac{1}{2}''$ thick.

Apply to the solution of this problem the principles explained for Fig. 224.

FIG. 227.   FIG. 226.   FIG. 224.

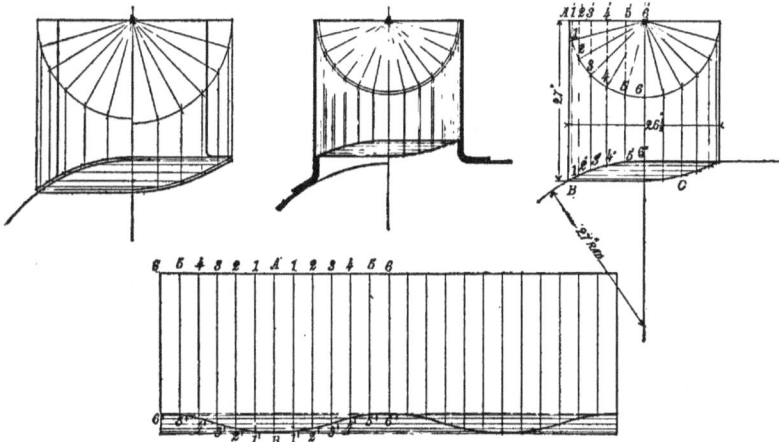

FIG. 225.

When your drawings are completed, compare them with Figs. 226 and 227, which are the projections required in the problem.

Letter or number the drawing and be prepared to explain how the different projections were found.

To draw the development of the top gusset-sheets of a locomotive wagon-top boiler of given dimensions.

First draw the longitudinal cross-section of the boiler to dimensions given by Fig. 228, using the scale of $1'' = 1$ ft.

Then at any convenient point on your paper draw a

straight line, and upon it lay off a distance AB $35\frac{1}{2}''$ long=
the straight part of the top of the gusset-sheet G, Fig. 225–228.
With center A and a radius=$27\frac{7}{8}''$ (the largest radius of the

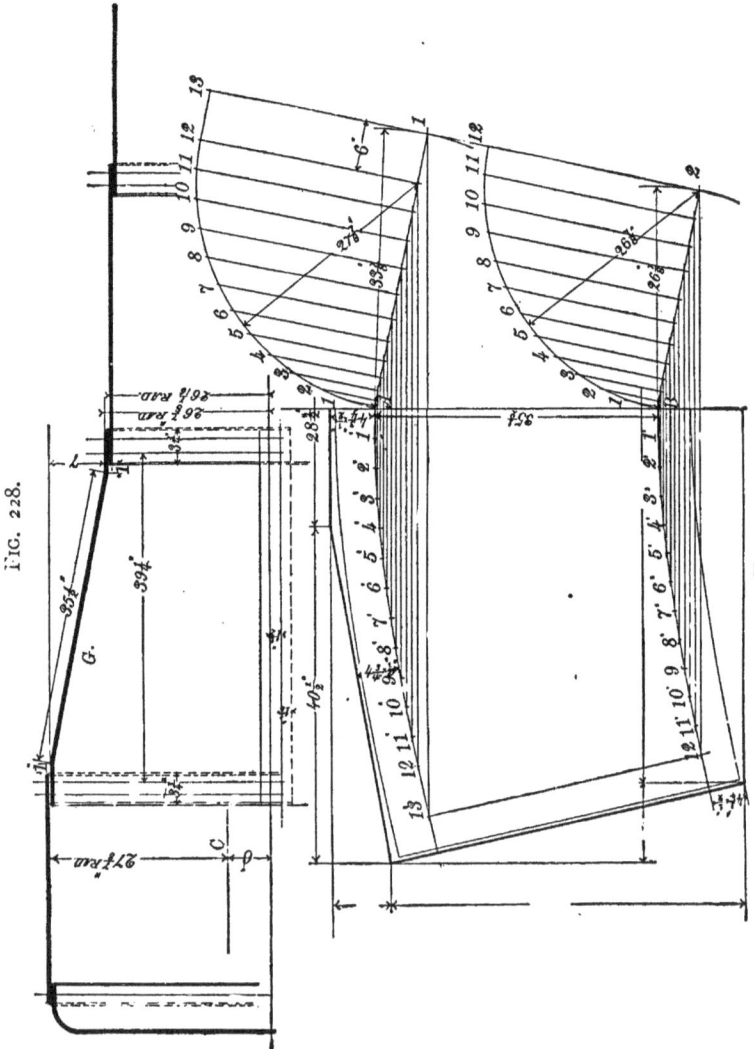

gusset)+$6''$ (the distance from the center of the boiler to the
center of the gusset C, Fig. 228)=$33\frac{7}{8}''$, draw arc 1.
   With center B and a radius=$26\frac{7}{8}$ (the smallest radius of

the gusset) draw arc 2. Tangent to these arcs draw the straight line 1, 2 extended, and through the points A and draw lines 1, A and 2, B perpendicular to 1, 2.

Take a point on the perpendicular 1, 2, 6″ from the point 1 as a center and through the point A draw an arc with a radius $= 27''\frac{7}{8}$.

With point 2 as a center and $2B$ as a radius $(26\frac{7}{8}'')$ draw an arc through B to meet the line 1, 2.

Divide both arcs into any number of parts, say 12, and through these divisions draw lines perpendicular to and intersecting $1A$ and $2B$, respectively. Through these intersections draw indefinite horizontals and on these horizontals step off the length of the arcs, with a distance $=$ one of the 12 divisions as follows:

On the first horizontals lay off the length of the arc $A1'$ and $B1' = A1$ and $B1$ respectively. Then from $1'$ lay off the same distance to $2'$ on the second horizontals, etc. Through these points draw curves $A13'$ and $B12'$. Join points $12'$ and $13'$ with a straight line. Then $AB12$, 13 will be the developed half of the straight part of the gusset.

On the two ends or front and back of the gusset we have now to add 1″ for clearance $+3\frac{3}{4}''$ for lap $+\frac{1}{2}''$ allowance for truing up the plates, total $= 5\frac{1}{4}''$. And to the sides $2\frac{5}{8}''$ for lap $+\frac{1}{2}''$ allowance for truing up, total $= 3\frac{1}{8}''$.

The outline of the developed sheet may now be drawn to include these dimensions with as little waste as possible, as shown by Fig. 229. Extreme accuracy is necessary in making this drawing, as the final dimensions must be found by measurement.

To draw the projections of a V-threaded screw and its nut of 3″ diameter and $\frac{3}{4}''$ pitch.

Begin by drawing the center line C, Fig. 230, and lay off on each side of it the radius of the screw $1\frac{1}{2}''$. Draw AB and $6D$. Draw $A6$ the bottom of the screw, and on AB step off the pitch $= \frac{3}{4}''$, beginning at the point A.

On line $6D$ from the point 6 lay off a distance $=$ half the pitch $= \frac{3}{8}''$, because when the point of the thread has completed half a revolution it will have risen perpendicularly a distance = half the pitch, viz., $\frac{3}{8}''$.

Then from the point $6''$ on $6D$ step off as many pitches as may be desired. From the points of the threads just found, draw the 30° triangle and T-square the V of the

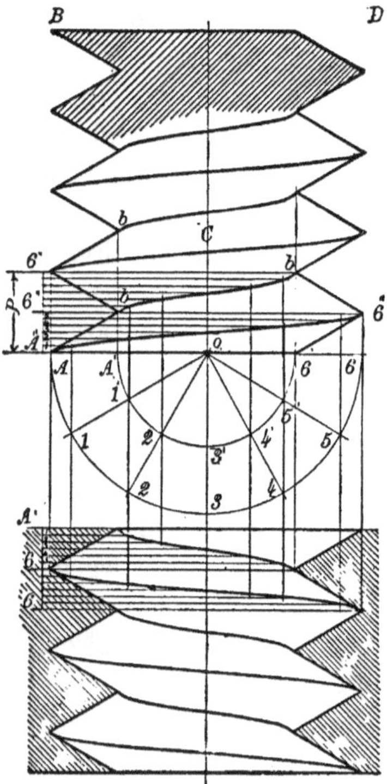

FIG. 230. 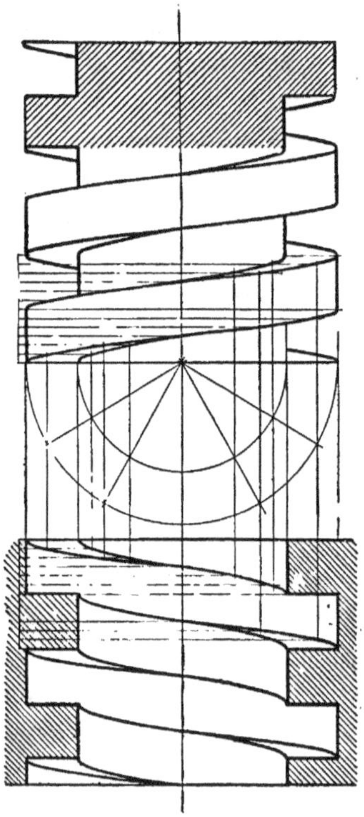　　　　　FIG. 231.

threads intersecting at the points $b..b..$ the bottom of the threads.

At the point O on line $A6$ draw two semicircles with radii = the top and bottom of the thread, respectively. Divide these into any number of equal parts and also the pitch P into the same number of equal parts. Through these divisions draw horizontals and perpendiculars intersecting each other in the points as shown by Fig. 230, which shows an elevation partly in section and a section of a nut to fit the screw.

Through the points of intersection draw the curves of the helices shown, using Irregular Curve, No. 13, Fig. 28.

FIG. 231. *To draw the projection of a square-threaded screw 3″ diameter and 1″ pitch and also a section of its nut.*

The method of construction is the same as for the last problem, and is illustrated by Fig. 231.

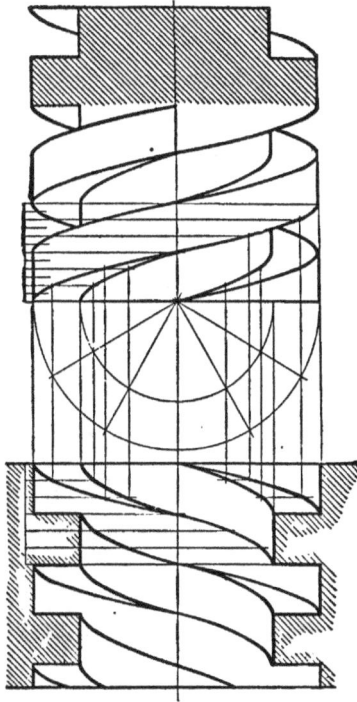

FIG. 232.

FIG. 232. *To draw the projections of a square double-threaded screw of 3″ diameter and 2″ pitch, and also a section of its nut.*

Proceed in the same manner as explained for the V-threaded screw, Fig. 230.

FIG. 233. *To draw the curve of intersection that is formed by a plane cutting an irregular surface of revolution.*

1st. Draw the complete outline of the connecting rod end partly shown in Fig. 233. Complete all three views.

2d. Divide line AB into any number of parts, say 14, preferably equal parts.

3d. With center D and radii equal to the distances from D to the several divisions on AB already determined, revolve those points to cut the line CD in a corresponding number of points.

4th. From the points on CD draw horizontals in narrow lines to the left of CD, cutting points in the $1\frac{5}{8}''$ radial curve G.

5th. From all these points in G drop perpendiculars to intersect horizontals from all the points in AB. Through the

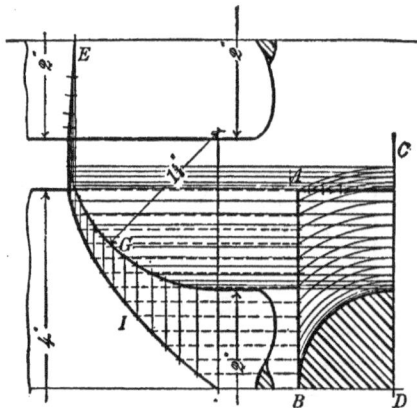

FIG. 233.

points of intersection draw the required curve of intersection I.

To find the curve in the plan at E.

1st. Divide the line AC into a convenient number of equal or unequal parts and revolve into CD.

2d. Through these points at C draw horizontals to intersect curve G extended and through points thus found erect perpendiculars.

3d. With the dividers take the several distances from CD along the line CA and lay them off from E to cut the corresponding perpendiculars from the extended part of the curve G. Draw the required curve through the points of intersection.

Figs. 233, 234, and 235 show examples of engine connecting rod ends where the curve I is formed by the inter-

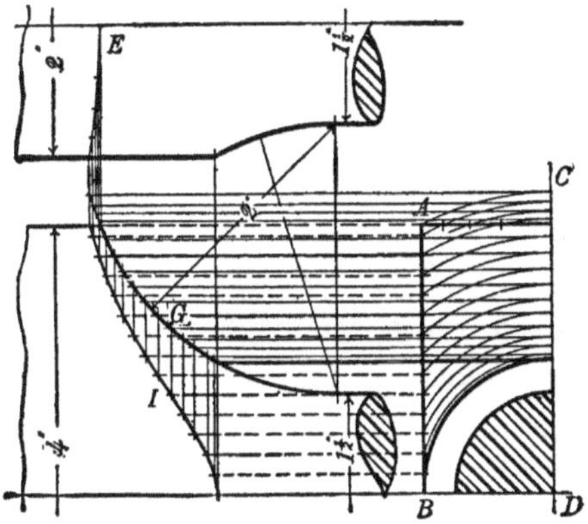

FIG. 234.

section of the flat stub end with the surface of revolution of
the turned part of the rod.

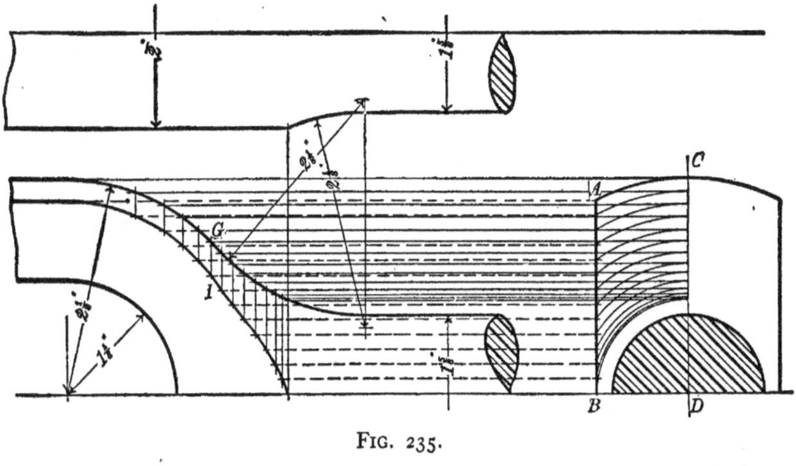

FIG. 235.

The description of the method used for finding the curve
of intersection I in Fig. 233 by intersecting planes will apply
equally as well to Figs. 234 and 235.

SHADE LINES, SHADES AND SHADOWS

Shade lines were used quite generally some years ago on commercial drawings. They improved the appearance of the drawing but the cost of applying them outweighed their value and so were abandoned except for special drawings.

The simple methods of application are included here for the benefit of those who nay desire to use them.

The *Shading* of the curved surfaces of machine parts is sometimes practiced on specially finished drawings, but on working drawings most employers will not allow shading because it takes too much time, and is not essential to a quick and correct reading of a drawing, especially if a system of shade lines is used.

. Some of the principles of shade lines and shading are given below, with a few problems illustrating their commonest applications.

The *shadows* which opaque objects cast on the planes of projection or on other objects are seldom or never shown on a working drawing, but are often projected on architecture and special drawings.

CONVENTIONS

The Source of Light is considered to be at an infinite distance from the object, therefore the Rays of Light will be represented by parallel lines.

The Source of Light is considered to be fixed, and the Point of Sight situated in front of the object and at an infinite distance from it, so that the *Visual Rays* are parallel to one another and perpendicular to the plane of projection.

Shade Lines divide illuminated surfaces from dark surfaces.

Dark surfaces are not necessarily to be defined by those surfaces which are darkened by the shadow cast by another part of the object, but by reason of their location in relation to the rays of light.

It is the general practice to shade-line the different projections of an object as if each projection was in the same plane, e.g., suppose a cube, Fig. 236, situated in space in the third angle, the point of sight in front of it, and the direction of the rays of light coinciding with the diagonal of the cube, as shown by Fig. 237. Then the edges $a^v b^v$, $b^v c^v$ will be shade lines, because they are the edges which separate the illuminated faces (the faces upon which fall the rays of light) from the shaded faces, as shown by Fig. 237.

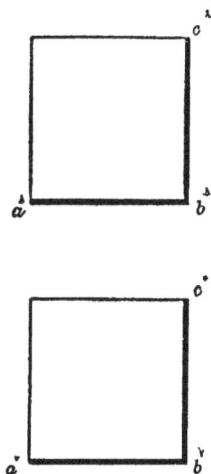

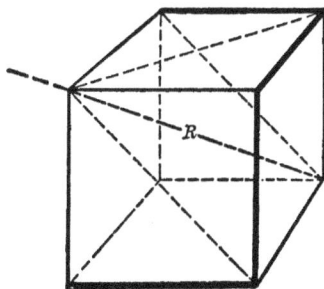

FIG. 236.   FIG. 237.

Now the source of light being fixed,. let the point of sight remain in the same position, and conceive the object to be revolved through the angle of 90° about a horizontal axis so that a plan of the top of the object is shown above the elevation, and as the projected rays of light falling in the direction of the diagonal of a cube make angles of 45° with the horizontal, then with the use of the 45° triangle we can easily determine that the lower and right-hand edges of the plan as well as of the elevation should be shade lines.

This practice then will be followed in this work, viz.:

Shade lines shall be applied to all *projections* of an object, considering the rays of light to fall upon each of them from the same direction.

Shade lines should have a *width* equal to 3 times that of the other outlines. · *Broken lines* should never be shade lines.

The outlines of *surfaces of revolution* should not be shade lines. The shade-lined figures which follow will assist in illustrating the above principles; they should be studied until understood.

SHADES

The *shade* of an object is that part of the surface from which light is excluded by the object.

The *line of shade* is the line separating the shaded from the illuminated part of an object, and is found where the rays of light are tangent to the object.

Brilliant Points. "When a ray of light falls upon a surface which turns it from its course and gives it another direction, the ray is said to be reflected. The ray as it falls upon the surface is called the incident ray, and after it leaves the surface the reflected ray. The point at which the reflection takes place is called the *point of incidence.*

" It is ascertained by experiment—

" (a) That the plane of the incident and reflected rays is always normal to the surface at the point of incidence;

" (b) That at the point of incidence the incident and reflected rays make equal angles with the tangent plane or normal line to the surface.

" If therefore we suppose a single luminous point and the light emanating from it to fall upon any surface and to be reflected to the eye, the point at which the reflection takes place is called the brilliant point. The brilliant point of a surface is, then, the point at which a ray of light and a line drawn to the eye make equal angles with the tangent plane or normal line—the plane of the two lines being normal to the surface."—Davies: *Shades and Shadows.*

Considering the rays of light to be parallel and the point of sight at an infinite distance, the brilliant point on the surface of a *sphere* is found as follows: Let A^vC^v and A^hC^h,

Fig. 238, be a ray of light and $A^v A^h$ a visual ray. Bisect
the angles contained between the ray of light and the visual
ray as follows: Revolve $A^v C^v$ about the axis A^v until it
becomes parallel to the horizontal plane at $A^v C_1^v$. At C_1^v
erect a perpendicular to intersect a horizontal through C^h
at C_1^h join $C_1^h L^h$ (L may be any convenient point on the line
of vision), bisect the angle $L^h A^h C_1^h$ with the line $A^h D^h$.
Join $C^h L^h$ and through the point D^h, draw a horizontal

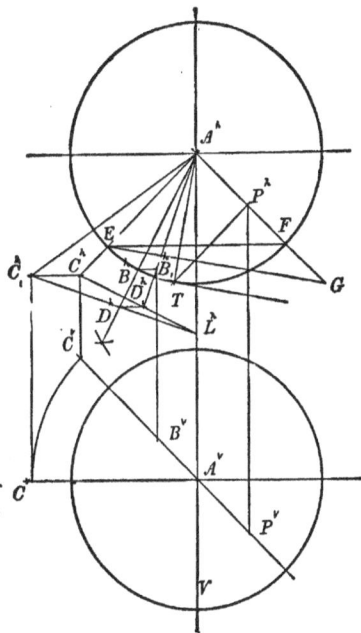

FIG. 238.

cutting $C^h L^h$ at D_1^h then $A^h D_1^h$ is the horizontal projection
of the bisecting line. A plane drawn perpendicular to this
bisecting line and tangent to the sphere touches the surface
at the points $B^v B_1^h$ where the bisecting lines pierce it.   Therefore $B^v B^h$ are the two projections of the brilliant point.

The point of shade can be found as follows:

Draw $A^h G$, Fig. 238, making an angle of $45°$ with a horizontal. Join the points E and F with a straight line EF.
Lay off on $A^h G$ a distance equal to EF, and join EG,

Parallel to *EG* draw a tangent to the sphere at the point *T*. Through *T* draw TP^h perpendicular to A^hG. From the point P^h drop a perpendicular P^r. P^r is the point of shade.

FIG. 239. *To shade the elevation of a sphere with graded arcs of circles.*

First find the brilliant point and the point of shade, and divide the radius 1. 2. into a suitable number of equal parts, and draw arcs of circles as shown by Fig. 239, grading them by moving the center a short distance on each side of the center of the sphere on the line B^h_2 and varying the length of the radii to obtain a grade of line that will give a proper shade to the sphere. It is desirable to use a horn center to protect the center of the figure.

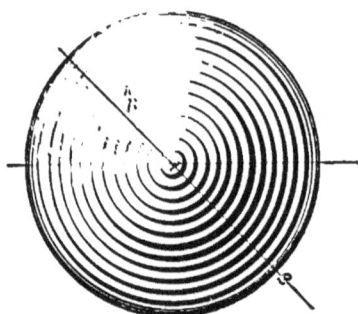 

FIG. 239.   FIG. 240.

FIG. 240 *shows the stippling method of shading the sphere.*

FIG. 241. *To shade a right cylinder with graded right lines.*

Find the line of light B' by the same method used to find the brilliant point on the sphere, except that the line of light is projected from the point B^h where the bisection line A^hD cuts the circle of the cylinder.

The line of shade is found where a plane of rays is tangent to the cylinder at S^r and S^h.

FIG. 242 *shows how the shading lines are graded from the line of shade to the line of light.*

It will be noticed that the lines grow a little narrower to the right of the line of shade on Fig. 242; this shows where the reflection of the rays of light partly illumine the outline of the cylinder.

FIG. 243. *To shade the concave surface of a section of a hollow cylinder.*

Find the element of light and grade the shading lines from it to both edges as shown by Fig. 243.

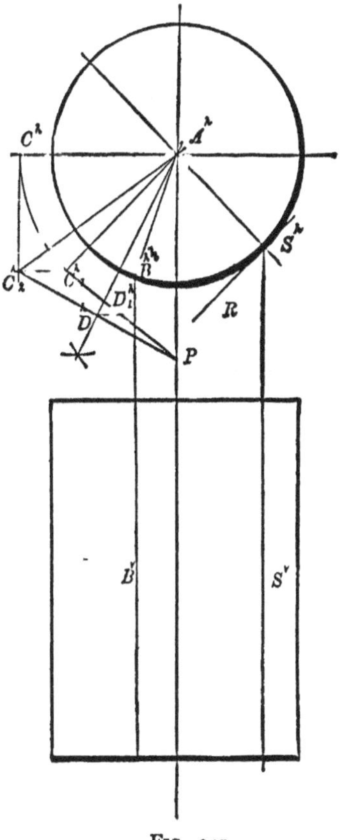

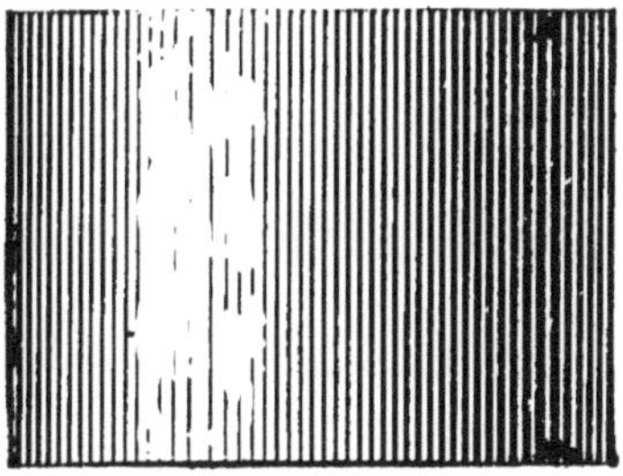

FIG. 242.

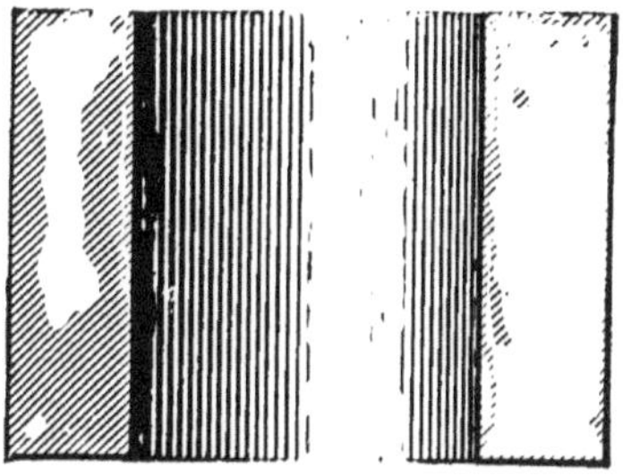

FIG. 243.

FIG. 241.

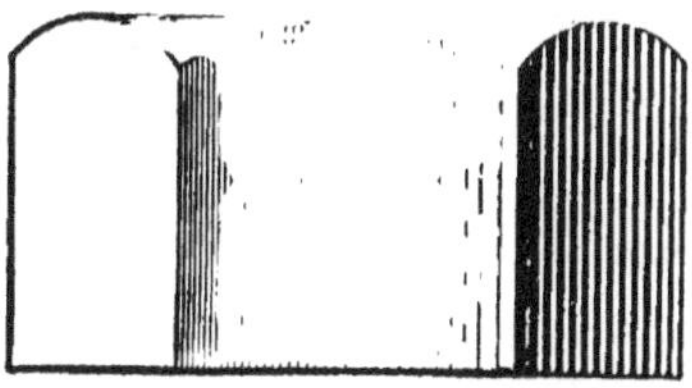

FIG. 244.

FIG. 244 *shows a conventional method of shading a hexagonal nut.*

FIG. 245. *To shade a right cone with graded right lines tapering toward the apex of the cone.*

Find the elements of light and shade as shown by Fig. 245, and draw the shading-lines as shown by Fig. 246, grading their width toward the light and tapering them toward the apex of the cone.

The mixed appearance of the lines near the apex of the cone on Fig. 246 can usually be avoided by letting each line dry before drawing another through it, or as some draftsmen do, stop the lines just before they touch./

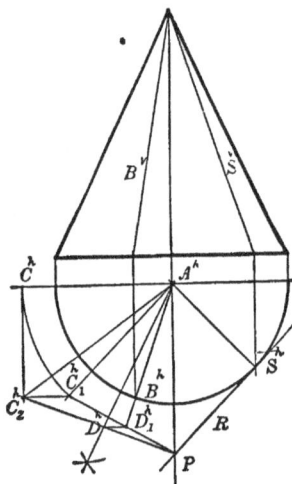

FIG. 245.

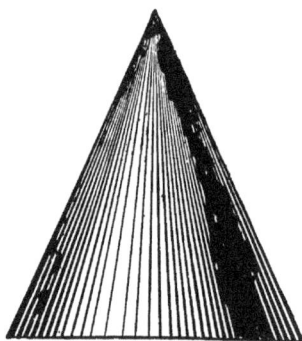

FIG. 246.

SHADOWS

, Let R, Fig. 247, be the direction of the rays of light and C an opaque body between the source of light and a surface S. The body C will prevent the rays from passing in that direction, and its outline will be projected at D on the surface S. D is the *shadow* of C.

The line which divides the illuminated portion of the surface S from the shadow D is called the *line of shadow*.

Shadow of a Point. If a line is drawn through a point in space in a direction opposite to the source of light, the point in which this line pierces the plane of projection is the shadow of the point on that plane.

To find the shadow on the H.P. of a point in space in the first dihedral angle·

Let A, Fig. 248, be the point in space, and R the direction of the ray of light; then $A_1{}^H$ is the shadow of the point A on H.P. and $A^H A_1{}^H$ is the horizontal projection and $A^V A_1{}^V$

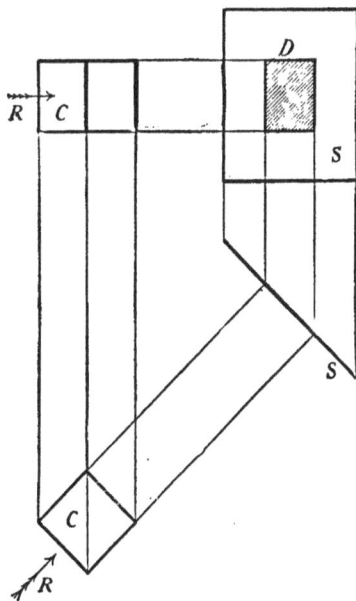

FIG. 247.

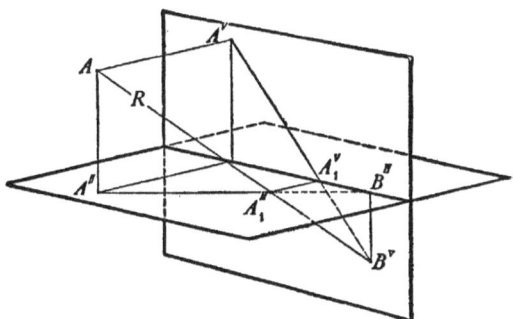

FIG. 248.

the vertical projection of R. B^V is the point where R pierces V when prolonged below H.P., and B^H is its horizontal projection in the G.L. The projections of R would then be $A^V B^V$.

The shadow of a point in V may be found in a similar manner.

Shadows of Right Lines. The shadow of a right line on a plane may be determined by finding the shadows of two of its points and joining these by a right line; e.g., the shadow of the line AB, Fig. 249, on H.P. is found as follows:

Through the points $A^V B^V$ draw the rays $A^V A_1^V$ and $B^V B_1^V$ to intersect the plane of projection in G.L. in the points A_1^V and B_1^V; from these points drop perpendiculars to meet rays drawn through A^H and B^H in the points A_1^H and B_1^H. A line drawn from A_1^H to B_1^H is the shadow of AB on H.P

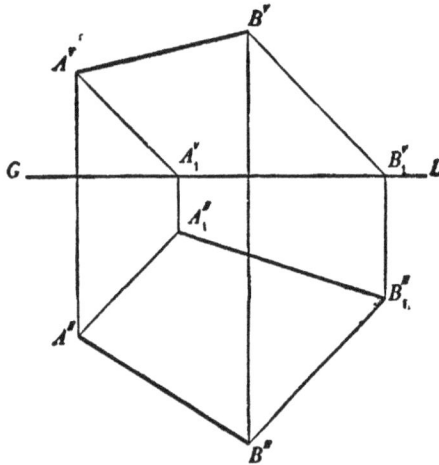

FIG. 249.

If a right line is parallel to the plane of projection its shadow will be parallel to the line itself.

If a line coincides with a ray of light, its shadow on any surface will be a point.

To find the shadow of a right line on V.P. and H.P.:

Let AB, Fig. 250 be the given line. Find the shadows points A and B by passing rays through each of their projections to make angles of 45° with G.L. The shadow of A^H on H.P. is found at A_1^H, and that of B^H at B_1^H, where the rays through these points intersect the H.P. The shadow of A^V on V.P. is found at $A_1{}^V$ and that of B^V at $B_1{}^V$, where the rays through these points intersect V.P. Join $A_1{}^H$ and $B_1{}^H$ with a straight line and we have the shadow of AB on H.P., and the shadow on V.P. is found in the same way by joining with a straight line the points $A_1{}^V$ and $B_1{}^V$.

That part of the shadow which falls on V.P. below G.L., and on H.P. above G.L., is called the *secondary* shadow, because it makes a second intersection, i.e., it is conceived to have passed through V.P. and made an intersection with H.P. behind V.P.

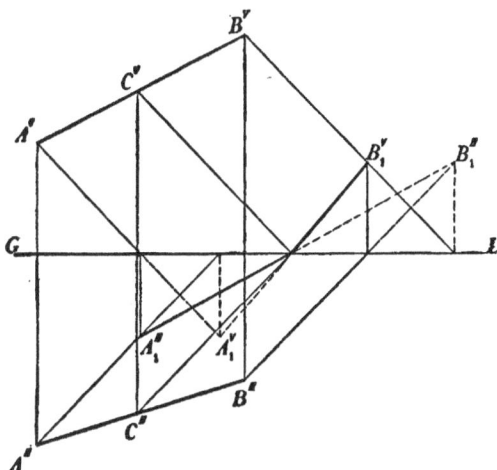

FIG. 250.

With the use of the secondary shadow problems like this are easier of solution.

ABCD, Fig. 251, is a square plate parallel to V.P.; find its shadow on H.P.

Through the points A^V, B^V, D^V, and $A^H C^H$, $B^H D^H$, draw rays making the angle of 45° (or any other angle which may be adopted) with G.L., and determine the shadows of these points as explained in Fig. 248. They will be found in the points $A_1{}^H B_1{}^H$, $C_1{}^H$, $D_1{}^H$. Join these points with right lines and they will form the line of shadow of the square plate on H.P.

FIG. 252.   *To find the shadow of a cube on V.P. with one face in V.P. and the other faces parallel or perpendicular to H.P.*

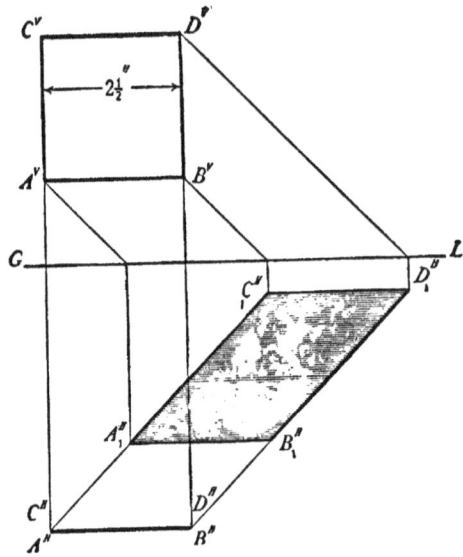

FIG. 251.

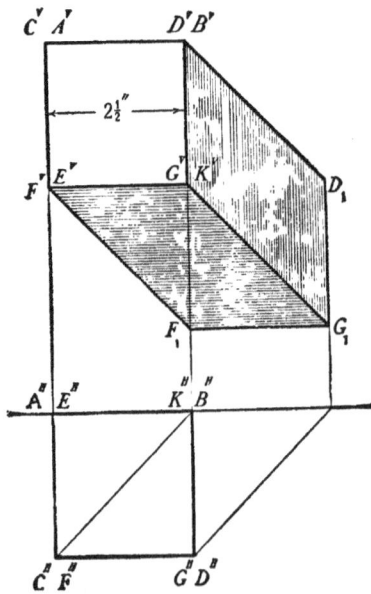

FIG. 252.

FIG. 252 *shows the cube in the given position. The line of shade is composed of edges EF, FG, GD, DB, and the edges AE and AB in V.P. which coincide with their shadows.*

The shadow of EF is E^VF_1, of FG is F_1G_1, of GD is G_1D_1, of DB is D_1B^V. The shadows of the edges AE and AB coincide with the lines. These shadows are found by the same rules used for finding the shadows of a line in Fig. 250. The line of shadow is $B^VD_1G_1F_1F^VE^VA^VD^V$. The visible line of shadow is $B^VD_1G_1F_1E^VC^VD^V$.

FIG. 253. *To find the shadow of a rectangular abacus on the face of a rectangular pillar.*

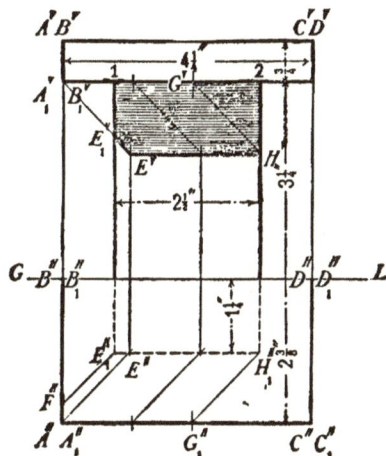

FIG. 253.

Assume the horizontal and vertical projections of the abacus and pillar to be as shown in Fig. 253.

The line of shade of the abacus is seen to be the edges $A_1{}^HB_1{}^H$ and $A_1{}^HC_1{}^H$. The plane of rays through edge $A_1{}^HB_1{}^H$ is perpendicular to V.P. and the line $A_1{}^VE^V$ is its vertical projection or trace; its horizontal trace is $A_1{}^HE^H$. The shadow on the left side face is vertically projected in the point $E_1{}^V$ where the plane of rays intersects that face. The ray through the point $A_1{}^H$ pierces the front face in the point E^H, which is the shadow of $A_1{}^H$, and $E_1{}^HE^H$, $E_1{}^Ve^V$ is the shadow of the part $F^HA_1{}^H$ on this face.

The line $A_1{}^H C_1{}^H$ is parallel to the front face, therefore its shadow on it will be parallel to itself and pass through E.

The visible line of shadow is now found to be $1 E_1{}^V E^V H^V 2$ 1.

FIG. 254. *Construct the shade of an upright hexagonal prism and its shadow on both planes.*

FIG. 254 shows the given prism with its line of shade

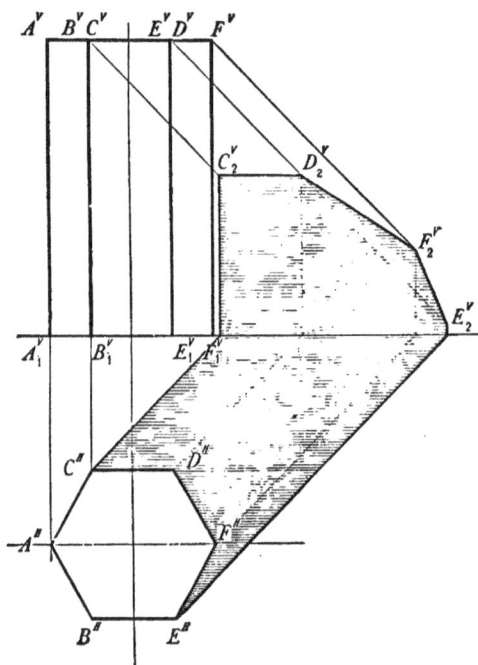

FIG. 254.

$A_1{}^V B_1{}^V E_1{}^V D^V F^V$ on the vertical projection, $C^H D^H F^H E^H$ on the horizontal projection, and its shadow on both planes.

FIG. 255. *Given a circular plate parallel to one coordinate plane: construct its shadow on the other plane.*

Let $A^V B^V C^V D^V$ and $A^H C^H$, Fig. 255, be the projections of the circular plate.

Circumscribe a square $E^V G^V$ about the circle; its shadow on H.P. will be the parallelogram $A^H G^H$, and the shadows of the points $A^V B^V C^V D^V$ are projected in the points $A_1{}^H B_1{}^H C_1{}^H D_1{}^H$.

The shadow of the inscribed circle is an ellipse tangent to the parallelogram at the points $A_1{}^H B_1{}^H C_1{}^H D_1{}^H$, with $B_1{}^H D_1{}^H$ and $A_1{}^H C_1{}^H$ as conjugate diameters.

The position and length of the axes of the ellipse of shadow may be found as follows:

Erect a perpendicular at the point C^V, making $G^V K^V$ equal to radius of the circle draw KOP; then KP is equal to the major and MK to the minor axis, and angle θ is twice the angle of the transverse axis with the horizontal conjugate diameter; i.e., KP is equal to 1, 2, MK to 3, 4, and 2, $O_1 C_1{}^H$, or angle θ, is equal to half KOC^V.

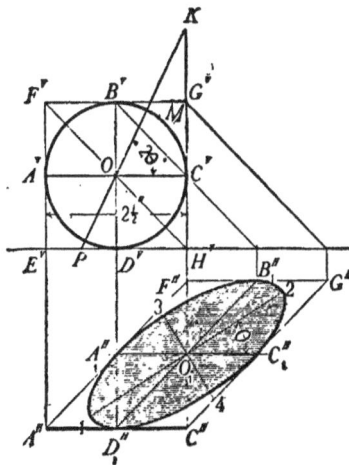

FIG. 255.

IG. 256. *Find the shade of a cylindrical column and abacus and the shadow of the abacus on the column.*

Let $A^V B^V C^V$ and $A^H B^H C^H$, Fig. 256, be the projections of the abacus, $D^H E^H F^H$ and $D^H D^V G^V F^H$ the projections of the column.

The line of shade on the column is found by passing two planes of rays tangent to the column perpendicular to H.P. and parallel to the horizontal projection of the ray of light. KL and E^H are the traces of these planes tangent to the column at the points L, and E^H and MN the visible line of deepest shade on the cylindrical column.

The deepest line of shade 1, 2 on the abacus is found in the same way.

The line of shadow on the column of that portion of the lower circumference of the abacus which is toward the source of light is found by passing vertical planes of rays, as 3, 4 to determine any number of points in the line, and joining these points by a line as shown in Fig. 256.

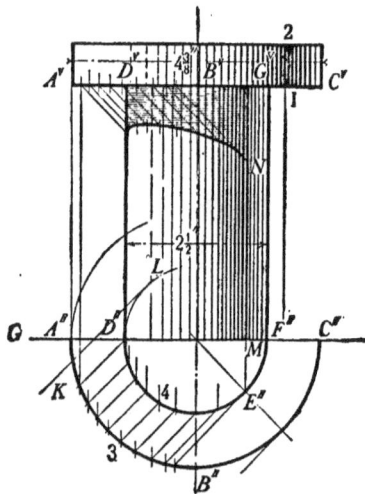

FIG. 256.

FIG. 257. *Find the shade of an oblique cone and its shadow on H.P*

Take the cone as given in Fig. 257 Pass two planes of rays tangent to the cone; their elements of contact will be the deepest lines of shade. To determine the elements of contact draw a ray through C^V; $C_1{}^H$ is its hor. trace. From $C_1{}^H$ draw lines tangent to the base at D and E; the lines of contact are CE and CD, and ECD is the line of shade.

The visible line of shade on H.P. is $E^H D^H$, and on V.P. $C^V E^V$. The shadow on H.P. is $E^H C_1{}^H D^H$.

Conic Sections. Fig. 258 shows a right circular cone cut by planes 1, 2, 3, and 4. Plane 1, perpendicular to the axis, cuts a *circle*. Plane 2 cuts the cone at an angle greater than that of the elements, and the section is an *ellipse*. Plane 3 cuts the cone at an angle equal to that of the elements and gives a *parabola*. Plane 4 cuts the cone at a smaller angle than that of the elements and the section is *hyperbola*. The

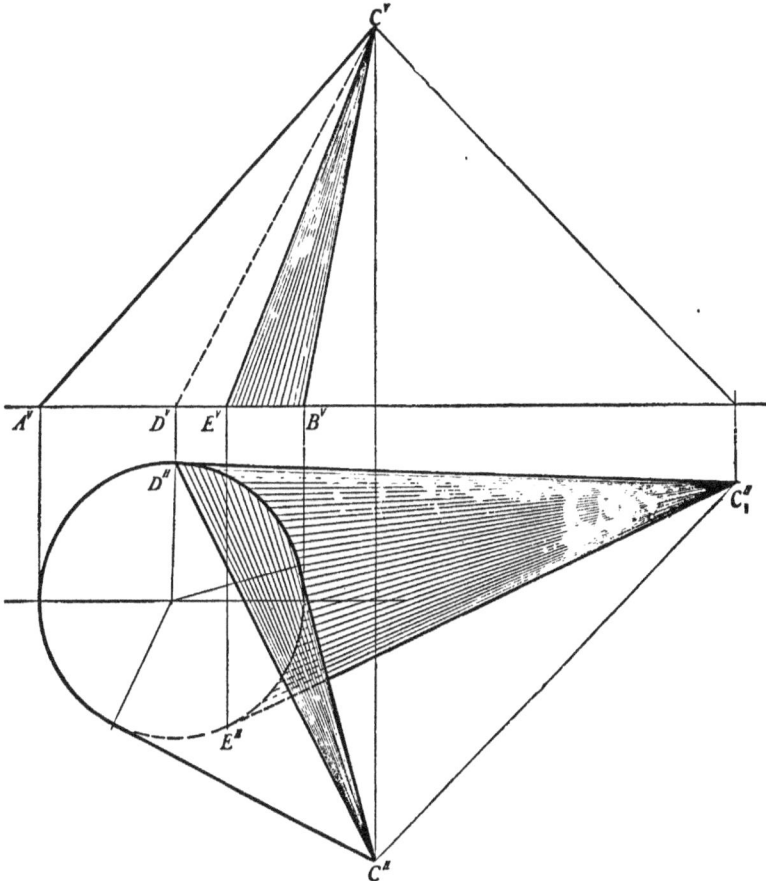

FIG. 257.

figure marked *A* is the elevation of the cone showing the cutting planes. *B* is the plane of the cone and the sections cut from it. *C* is the profile. *D* is the developments or true sizes of the truncated cone below planes 1, 2, 3, and 4. *E* is the true section of the hyperbola. *F* the true section of parabola, and *G* the true section of the ellipse.

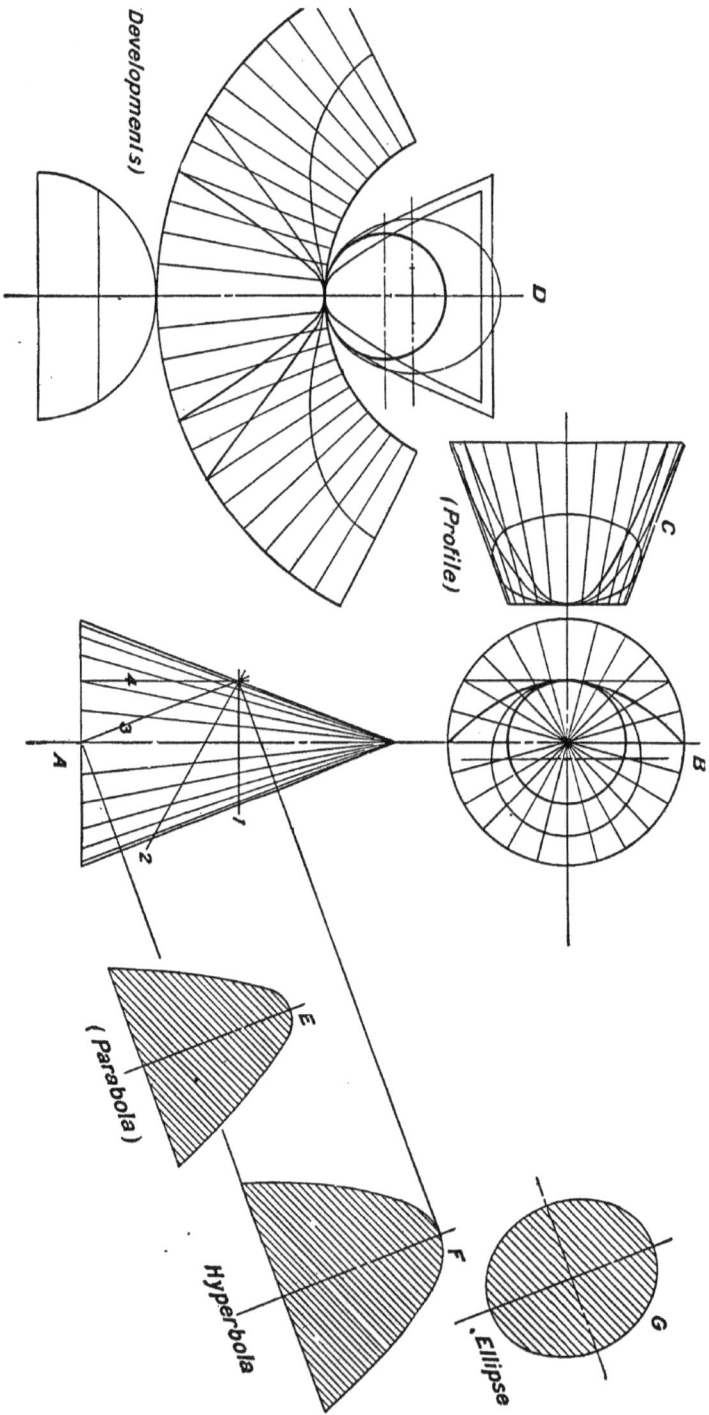

FIG. 258.

PRESENT PRACTICE IN DRAFTING ROOM CONVENTIONS AND METHODS IN MAKING PRACTICAL WORKING DRAWINGS.

SUMMARY REPORT OF AN INVESTIGATION MADE BY THE WRITER WITH THE AUTHORITY OF THE ARMOUR INSTITUTE OF TECHNOLOGY, CHICAGO, ILL., INTO THE PRESENT PRACTICE OF THE LEADING DRAFTSMEN IN THE UNITED STATES, IN THE USE OF STANDARD CONVENTIONS AND METHODS WHEN MAKING COMMERCIAL WORKING DRAWINGS.

A circular letter accompanied by a list of thirty-five questions was submitted to two hundred leading firms in the United States, embracing nearly all kinds of engineering practice.

The returns have been exceedingly gratifying, and especially so has been the spirit with which the "Questions" have been received and answered.

Many requests have been received from chief draftsmen for a copy of the returns.

The questions submitted and the answers received are given somewhat in detail below.

Q. 1. Do you place complete information for the shop on the *pencil drawing*, such as all dimensions, notes, title, bill of material, scale, etc.?

Complete information is placed on drawing before tracing.... 57
Complete information is placed on tracing only........... 42
Principal dimensions and title only on pencil drawing....... 2
Draw directly on bond paper........................... 10
Did not answer this question.......................... 10
Sometimes.. 7

Reasons given for making the pencil drawing complete:

To arrange notes. To save time. The tracing is not usually made by the draftsman who makes the pencil drawing.

Q. 2. Do you ever ink the pencil drawing?

Never ink the pencil drawing.............................. 91
Generally ink the pencil drawing........................ 7
Sometimes ink the pencil drawing....................... 8
Sometimes ink the pencil drawing and shellac it for shop use. 1
Use bond paper.. 10
Make pencil drawings on dull side of tracing cloth.......... 2
Ink center lines of assembly drawing..................... 1
Ink center lines of pencil drawings in red................. 2

Q. 3. Do you trace on cloth and blue print?

Always trace on cloth and blue print..................... 102
Blue print from bond paper............................. 10
Blue print from bond paper occasionally.................. 1
Sometimes make " Vandyke " prints for shop use.......... 1
Sometimes use paper drawings in shop for jigs and fixtures... 1

Q. 4. Do you use blue prints entirely in the shop?

Use blue prints altogether in shop....................... 105
Sometimes use pencil drawings or sketch.................. 21
Sometimes use sketches made with copying ink............ 1
Sometimes use prints from " Vandyke ".................. 1
Use white prints mounted on cardboard and varnished...... 1
Use blue prints mounted on cardboard.................... 1
Use sketches for rush work.............................. 1

Q. 5. When tracing do you use uniform wide object lines? Ever use shade lines?

Use uniform, thick object lines. Never use shade lines...... 100
Sometimes use shade lines............................... 21
Use shade lines on small details......................... 5
Always use shade lines.................................. 14
Experts in the use of shade lines may do so to make drawings clear.. 1
Shade rounded parts.................................... 1

Q. 6. What kind of a center line do you use?

Long dash, very narrow, and dot, thus: —— · ——	42
Long dash and two dots, —— ·· ——	29
Very fine continuous line, ——————	19
Very fine dash line, long dashes, —— ——	8
Long dash and dot in red, —— · ——	3
Continuous fine red line, ——————	8
Long dash and three dots, ——— ··· ———	1
Long dash and two dots, thus: —— ·· ——	1

Q. 7. What kind of dimension line do you use?

Continuous fine line, broken only for dimension——	52
Fine long dash line, ——————	32
Fine long dash line and dot, —— · ——	13
Fine continuous red line, ——————	8
Fine continuous blue line, ——————	4
Fine continuous green line, ——————	1
Same character of line as center line,...................	2
Dotted line, - - - - - - - - - - - - - - - - -	1
Long dash and two dots, —— ·· ——	2
Heavy broken line, —— —— ——	1

Q. 8. What style of lettering do you use? Sloping? Vertical?
Free-hand? All capitals of uniform height? or capitals
and lower case?

Free-hand sloping................................	52
Free-hand vertical................................	45
Free-hand capitals, Gothic, uniform height..............	61
Free-hand capitals, and lower case....................	40
All caps, initials slightly higher.....................	5
Lettering left to option of draftsman..................	2
Mechanical lettering, all caps.......................	3
Not particular, the neatest the draftsman can make free hand	4
Mechanical lettering, all caps, sloping.................	2
Give great latitude in lettering, only insist it be bold and neat	1
Roman, caps and lower case, free hand.................	2
Large letters $\frac{3}{16}$ths, small $\frac{3}{32}$ds and $\frac{1}{8}$th..............	2

Q. 9. Are your titles and bills of material printed or lettered by
    hand?

Lettered by hand...................................... 79
Standard titles printed and filled in by hand.............. 12
Bill of material table printed and lettered by hand......... 12
Lettered by hand, contemplate having them printed........ 1
B. of M. typewritten on separate sheet and blue printed..... 8
Titles partly printed and filled in by hand................ 8
Use rubber stamp for standard title, fill in by hand......... 6
Standard title, bill of material lithographed on tracing cloth 8

Q. 10. Do you use a border line on drawings?

Always use border lines.............................. 97
Never use border lines................................ 13
Use border lines on foundation plans, to send out.......... 1
No border lines on detail drawings...................... 1
Intend to discontinue the use of border lines.............. 1
Border lines used only on design drawings................ 1
Only on drawings to be mounted on cardboard......... ... 1
Only used for trimming blue print...................... 2
On assembly drawings only............................ 1
Width of margins reported: 1", ½". ⅜", ¼", and ⅛".

Q. 11. When hatch-lining sections, do you use uniform or
    symbolic hatch lines?

Standard symbolic lines.............................. 59
Uniform hatch lines for all materials.................... 44
Shade section part with 4H pencil and note name of material 4
Symbolic hatch lines and add name of material........... 3
Uniform hatch lines for metal only...................... 1
Uniform on details, symbolic on assembly drawings......... 5
Pencil hatch on tracings and note material other than cast iron 1
Uniform hatch lines, sometimes solid shading............. 1
No uniform system.................................. 1
Sections tinted with water colors representing the metals.... 1

Q. 12. Is the pencil drawing preserved?   Is the tracing stored
or do you make "Vandyke" prints for storing away?

Store tracings only	96
Pencil drawings preserved for a time	30
Pencil drawings preserved	3
White prints made and bound for reference	1
Tracings kept in office for reference, blue prints stored	9
" Vandyke " prints stored	1
Use " Vandyke " as substitute for tracing	2
Arrangement drawings preserved, detail drawings destroyed after job is completed. Pencil drawings used for gasket paper	1
Original pencil drawing inked and stored	1
Assembly drawings and layouts preserved	4
Patent office drawings preserved	1
Tried " Vandyke " but found it unserviceable, tearing easily	1

Q. 13. Do you use 6H grade of pencil for pencil drawings or
what?

6H	73
4H, mostly for figures and letters	52
5H	16
Ranging from 2H to 8H	53

Q. 14. Do you use plain orthographic projection for free-hand
sketches?   Ever use perspective or isometrical drawing
for sketches?

Plane orthographic 3d angle projection	99
Insometrical drawing for sketches	25
Perspective for sketches	1
Isometric for piping layouts and similar work	8
Perspective and isometric for catalogue work	2
Isometric sometimes	6
Never use free-hand sketches	6

One says, "When we run into other than orthographic, men are too
timid and not sure of themselves.   In perspective drawings when work is
cylindrical, workmen get mixed up on center lines.

Q. 15. What sizes of sheets do you use for drawings?

9″×12″... 13
12″×18″... 16
18″×24″... 20
24″×36″... 19

There seems to be little uniformity in the sizes of shop drawings, about 67 firms reporting different combinations. A few have no system but simply make the size of sheet to suit the object to be drawn.

Q. 16. Do you use red ink on tracings?

Never use red ink on tracings........................... 57
Recently discarded the use of red ink.................... 2
Use red ink for pattern figures......................... 1
Use red ink for center and dimension lines............... 8
Use red ink for check marks............................ 1
Use red ink for existing work on studies................. 1
Use red ink sometimes................................. 2
Use red ink on occasions when it is desired to show old work
    in red and new work in black (use carmine)............. 1
Use carmine for brick................................. 1

Qs. 17 and 27. How indicate finished surfaces on drawings?
    When finished all over? When "file finished," ground
    planed, bored, drilled, etc.?

Finished surfaces indicated as in Fig. 1.................. 65
Finished surfaces indicated as in Fig. 2.................. 16
Finished surfaces indicated as in Fig. 3.................. 8
Finished surfaces indicated as in Fig. 4.................. 2
Finished surfaces indicated as in Fig. 5.................. 2
Bound the surfaces with red lines....................... 2
Bound the surfaces with dotted lines.................... 2
Name the finish by note in full......................... 68
Do not specify machinery method....................... 6

(See drawing.)

Q. 18. Do you use horizontal or sloping lines for convention in screw threads?

Sloping lines, see Fig. 6.................................. 94
Horizontal lines, see Fig. 7.............................. 12

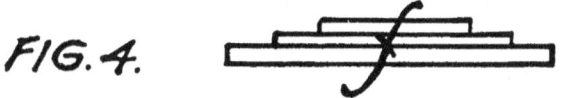

FIG. 1.

FIG. 2.

FIG. 3.

FIG. 4.

FIG. 5.

FIG. 6.   FIG. 7.   FIG. 8.   FIG. 9.   FIG. 10.

Horizontal lines, see Fig. 8............................. 13
Both.. 7
Neither, but as shown in Fig. 9......................... 1
Neither, but as shown in Fig. 10........................ 1

Q. 19. When a large surface is in section do you hatch-line around the edges only?

Hatch-line edges only.................................... 62
Sometimes... 3
Hatch section all over.................................. 54
Do not use hatch lines; shade the whole surface with 4H pencil 3
Usually show a broken surface line..................... 1

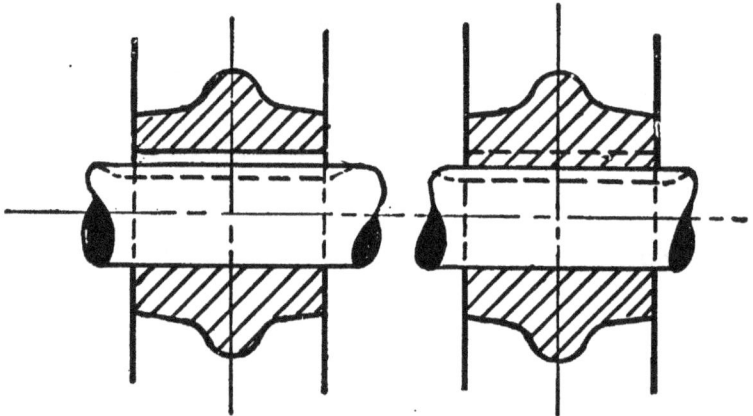

FIG.11.   FIG.12.

Q. 20. Do you section keyways in hubs or show by invisible lines?

Section keyways as shown in Fig. 11...................... 73
Show keyway by invisible lines, see Fig. 12............... 40
Keyways in hubs left blank............................. 1

Q. 21. In dimensioning do you prefer to place the dimension upon the piece or outside of it?

Outside whenever possible.............................. 92
Upon the piece.. 13
Both, according to size and shape of part................ 19
No rule.... 1

Commenting on placing dimensions outside of piece one says, "It entails less confusion to workman." Another says: "So as to make detail stand out."

Q. 22. Do you use feet and inches over 24 inches?

Yes...	69
Use feet and inches over 36"...........................	4
Use feet and inches over 24" on foundations and outlines....	2
Use feet and inches over 48"...........................	6
All inches...	21
For pulleys use inches up to 48".......................	1
Inches up to 10 feet...................................	2
Start feet 24" thus: 2⊥0"............................	2
Usually, but not always................................	2
Yes, except pitch diameters of gears, which are all given in inches..	2
Yes, except in boiler and sheet iron work..................	3
Use feet and inches over 12"...........................	6
Inches up to 100".....................................	3
Inches up to 60"......................................	1

Q. 23. How do you indicate feet and inches? Thus 2 ft. 4", or thus 2⊥4"?

2⊥4"—97, 2$^{\text{FT.}}$ 4"—5, 2 FT. 4"—2, 2ft. 4"—13. Both 2ft. 4" and 2-4"—1, 2FT. 4 IN.—1, 2' 4"—8, 2 ⸜ 4 ⸝—1.

Q. 24. Do you dimension the same part on more than one view?

One view..	94
More than one view as check...........................	46

Q. 25. When several parts of a drawing are identical would the dimensioning of one part suffice for all, or would you repeat the dimension on each part?

One part only...	82
Would repeat or indicate by note.......................	39
"Left to judgment of draftsman".......................	1

" When it is evident that several parts are identical the dimensioning of one part would suffice, ' Would never leave room for doubt.' "

Q. 26. Do you write R for radius or RAD.? D. for diameter or DIA.?

RAD.....35	Rad....47	R......32	rad..... 1	r....... 3
DIA.....41	Dia.....48	D......15	d....... 3	dia..... 4
	DIAM........ 1	Diam..... 3	diam...... 5	

Do not use R. or RAD., dimension only.................... 1

Q. 28. Do you always give number of threads per inch? When you do how are they indicated?

Only give number of threads when not standard........ 67

All others always indicate number of threads in a great variety of ways. A few of the different styles of noting the threads are given below:

¾″—10 Thr. 5THDS. PER 1″. 8thds. 4 threads per inch. Mach. Screw 10–24, 1¼″ XII, 16 P. RH. Vth. U. S. S. XVIII, 1″–8– U. S. S. 1″ TAP, 8 PITCH, 3 TH'D R. H. SQ. DOUBLE, 5″–18 THDS. R. H. OWN ST'D 10 thds. per inch. For pipe tap thus, ½″ P.T., etc., etc.

Q. 29. How do you "Mark" a piece to indicate on the bill of material?

Number it on drawing and put a circle around it........... 34
By name or letter.................................... 35
By patron number.................................... 2
By symbol and number................................ 14
Castings, I, II, III, Forgings, 1, 2, 3.

Q. 30. When a working drawing is fully dimensioned why should the scale be placed on the drawing?

For convenience of drafting room...................... 25
Check against errors................................. 11
Not necessary....................................... 18
Scale not placed on shop drawings..................... 18
For convenience in calculations and planimeter work....... 1
To give an idea of over-all dimensions when these are not given. "We never saw a drawing so fully dimensioned as to warrant leaving off the scale"..................... 2

"If a drawing is to scale the scale should be on the drawing, whether it is needed or not."

"It gives every one interested a better conception of the proportions of the piece, and there are frequently portions of a design which do not require a dimension for the shop to work to, and which it is interesting to scale from an engineering point of view."

"To get approximate dimensions not given on drawing."

"Impractical to dimension all measurements for all classes of work."

"Scale will tell at a glance, dimensions would have to be scaled."

"To obtain an idea of relative size of parts without scaling the drawings."

" To sketch on clearance." " To proportion changes." " When erecting to measure over-all sizes."

" In case a dimension has been left off, the scale will help out."

" This is a question of opinion; some will not have the scale, others insist on it." " We always give the scale."

" It is an immense help and time saver in the drawing room."

" Generally no reason. In our work we combine standard apparatus by 'fudge' tracing, and it is convenient to know scale so all parts will surely be to same scale."

" In discussing alterations, additions, clearances, etc., it is convenient to know the scale instantly."

" For convenience in drafting room. We often put an arbitrary scale on with a reference letter indicating scale to draftsman."

" To give toolmaker an idea of the size of the finished piece."

" As an aid to the eye in reading."

Above are some of the reasons given for placing the scale on the drawing. Below are given a few of the reasons why some do not place the scale on the drawing.

" Scale should never be used in shop," says one.

" Not necessary. Sometimes drawing is made out of scale."

" Not advisable, on account of workmen getting into the habit of working to scale instead of to the figures."

" Know of no good reason at all."

" Believe it best to leave scale off."

" Should not. Drawing should never be scaled."

" Know of no good reason why it should be."

" Should not be given on drawing."

" Do not object if left off, not needed."

Q. 31. Do you use the glazed or dull side of tracing cloth?

Dull side......66   Glazed side..32   Both.......4

" Dull side, because it lies flat better in drawers."

" Dull side, so that changes which may be necessary while work is under construction, can be made easily in pencil and later in ink."

" Dull side so tracings may be checked in pencil."

" It prevents curling."

" Both, although the glazed side when traced on lies better in the drawer."

" We use cloth glazed on both sides, work on convex side, so that shrinkage of ink will eliminate camber."

" Dull, except for U. S. Government, who requires the glazed side to be used."

Q. 32. How do you place pattern numbers on castings?

> Pattern number with symbol or letter is placed on or near the
> piece, e.g., PATT.–D–478–C........................ 36

This question was not happily stated: most answers gave "raised letters cast on," while the question like all the others refers to the marking of the drawing.

Q. 33. How do you note changes on a drawing?

> On tracing with date................................. 32
> New tracing and new number........................... 17
> Put a circle around old figure and write new figure beside it
>   with date....................................... 8
> Make new tracing.................................... 5
> Red ink with date................................... 8
> Use rubber stamp "Revised " with date, and indicate changes
>   on record print................................. 28
> Use change card system.............................. 1
> Special forms for purpose. Change made in a book with
>   date. New prints made to replace. In place at title
>   with draftsman's initials and date............... 8

Q. 34. Do you place dimensions to read from bottom and right hand, or all to read from bottom, or how?

> Bottom and right hand....103   From bottom only........ 2
> No fixed rule...................................... 2
> From R to L and bottom to top....................... 1

Q. 35. Do you always make a table to contain the bill of material?

> Yes...........49 No...........25 Not always.... 5
> Usually.................. 1 Use separate bill....... 32
> Bills on general drawings only. On details number is marked on piece.
> " No, but it is advisable to do so." " Have abandoned that system."

INDEX

A

H

I

L

M

N

www.ingramcontent.com/pod-product-compliance
Lightning Source LLC
Chambersburg PA
CBHW021527210326
41599CB00012B/1409